geologic time

THE PRENTICE-HALL FOUNDATIONS OF EARTH SCIENCE SERIES
A. Lee McAlester, Editor

STRUCTURE OF THE EARTH

S. P. Clark, Jr.

EARTH MATERIALS

W. G. Ernst

THE SURFACE OF THE EARTH

A. L. Bloom

EARTH RESOURCES, 2nd ed.

B. J. Skinner

GEOLOGIC TIME, 2nd ed.

D. L. Eicher

ANCIENT ENVIRONMENTS

L. F. Laporte

THE HISTORY OF THE EARTH'S CRUST*

A. L. McAlester and D. L. Eicher

THE HISTORY OF LIFE

A. L. McAlester

OCEANS

K. K. Turekian

MAN AND THE OCEAN

B. J. Skinner and K. K. Turekian

ATMOSPHERES

R. M. Goody and J. C. G. Walker

WEATHER

L. J. Battan

THE SOLAR SYSTEM*

J. A. Wood

*In preparation

geologic time

second edition

DON L. EICHER
University of Colorado

PRENTICE-HALL, INC., *Englewood Cliffs, New Jersey*

Library of Congress Cataloging in Publication Data

EICHER, DON L
    Geologic time.

    (The Prentice-Hall foundations of earth science
series)
        Bibliography: p. 141
        Includes index.
        1. Geological time.   2. Geology, Stratigraphic.
I. Title.
QE508.E38 1976   551.7'01   75-25824
ISBN   0-13-352492-2
ISBN   0-13-352484-1 pbk.

10   9   8   7   6   5   4   3   2   1

Printed in the United States of America

PRENTICE-HALL INTERNATIONAL, INC., *London*

PRENTICE-HALL OF AUSTRALIA PTY. LIMITED, *Sydney*

PRENTICE-HALL OF CANADA, LTD., *Toronto*

PRENTICE-HALL OF INDIA PRIVATE LIMITED, *New Delhi*

PRENTICE-HALL OF JAPAN, INC., *Tokyo*

PRENTICE-HALL OF SOUTHEAST ASIA PTE. LTD., *Singapore*

contents

three

four

five

six

growth
of the concept

THE MEDIEVAL MYTH

Hendrik Van Loon once measured the scope of time in the following tale:

> High up in the North in the land called Svithjod, there stands a rock. It is a hundred miles high and a hundred miles wide. Once every thousand years a little bird comes to this rock to sharpen its beak.
>
> When the rock has thus been worn away, then a single day of eternity will have gone by.*

Since man first began writing his thoughts down, he has been persistently concerned about his own place in eternity. Until the writings of James Hutton in 1788, however, the concept of almost limitless time was reserved chiefly for man alone, and the Earth was viewed in a strictly temporal framework. In man-centered medieval thought, the Earth constituted a closed system, with a beginning not long ago, and an ultimate end not far in the future.

Christians of the prescientific era viewed their Earth as a massive, sluggish, totally immobile object at the center of the universe, beyond which lay the pure, ethereal realm, free from all blemish and corruption. This heavenly realm included, at a modest distance, the airy and weightless Sun, Moon, and planets, and an all-encompassing celestial sphere that contained all the stars. Everything,

---

* From *The Story of Mankind* by Hendrick Van Loon. By permission of Liverright, Publishers, N.Y. Copyright (©) 1951, by Liverright Publishing Corp.

FIG. 1-1 The medieval universe, with the Earth stationary at the center, was but a few thousand miles across. The medieval concept of time was similarly limited. (From Petrus Apianus' *Cosmographie*, 1551.)

of course, revolved about the Earth once each day. This concept of confined space is well illustrated by the old woodcut reproduced in Fig. 1-1. The confining nature of medieval time is closely parallel, but is not so easily illustrated. For this very reason the concept of sharply limited space is more easily discredited.

Early in the seventeenth century, Kepler (1571–1630) and Galileo (1564–1642), armed with both a healthy disrespect for the scientific authoritarianism of the day and the newly invented telescope, separately championed the Sun-centered celestial system envisioned earlier by Copernicus (1473–1543), and laid to rest the old idea of greatly confined space surrounding a motionless Earth. The Earth-centered concept died slowly and hard, but after the mid-seventeenth century it yielded to the modern concept of a dynamic, spinning Earth with vast if not infinite space extending outward from it in all directions.

Time was a more difficult matter. Because it is intangible, the initial perception of its dimensions demanded inspired insight, which the seventeenth-century naturalists were unable to muster; so the medieval view of a very short span of Earthly time remained. Christian scholars of the day generally assumed that the age of the Earth was about 6,000 years, a figure based on literal acceptance of the ancient Hebrew writings. Within the brief length of time thus allowed, it was unimaginable that splashing rain, ocean surf, and imperceptible crustal movements could have played a significant role in shaping the Earth's surface. Instead, its features, particularly the large features, seemed to require a special episode of beginning when the formative processes were violent catastrophes.

JAMES HUTTON: THE CONCEPT OF
UNIFORMITARIANISM

Near the end of the eighteenth century, James Hutton challenged this remnant of medieval thought. By then, most scientific inquiry was reasonably sophisticated, and was based on an intense desire for mathematical order. The idea of great catastrophes that temporarily override natural law had come to fit poorly in an otherwise rational universe.

Hutton's views were straightforward. A remarkably perceptive observer, he thought that he recognized in the rocks of his native Scotland the results of processes presently going on at the Earth's surface, processes such as erosion, deposition, and volcanic activity. Given enough time, the present rates of activity would be sufficient to produce all features of the rocks, and all of their observable relationships and configurations. Where the catastrophist viewed the Earth's surface as a short-lived remnant of a single tumultuous splash of creation, or perhaps a series of such tumultuous events, Hutton viewed it as a nearly eternal machine in which internal dynamic forces created stresses that, in the course of time, elevated new lands from the ocean bed even as other exposed surfaces were being eroded. Hutton saw no evidence of the universal flood championed in one form or another by most catastrophists, only signs of slow subsidence of the Earth's surface in some places and its renewed uplift in others. To him, the dynamic, spinning Earth of the astronomers had, in addition, a dynamic surface and a dynamic interior. Said Hutton, "From the top of the mountain to the shore of the sea . . . everything is in a state of change."* Through erosion, the Earth's surface locally deteriorates, but by rock-forming processes, it sumultaneously builds itself up elsewhere. The Earth, he said, "has a state of growth and augmentation; it has another state, which is that of diminution and decay. This world is thus destroyed in one part, but it is renewed in another."

Unlike his predecessors, Hutton always carefully cited verifiable observations. In arguing that mountains are sculptured and ultimately leveled by weathering and stream erosion, and that their fragments are carried to the sea by processes exactly like those acting presently, Hutton said, "We have a chain of facts which clearly demonstrates . . . that the materials of the wasted mountains have traveled through the rivers" and "There is not one step in all this progress . . . that is not to be actually perceived." Then he summed up, "What more can we require? *Nothing but time.*" And viewed in the light of Hutton's concepts of change through existing causes, the Earth appeared to be the product of almost limitless time. Man, said Hutton, has before him today all of the principles "from whence he may reason back into the boundless mass of time already elapsed." This point of view soon came to be called "uniformitarianism."

---

* From *Theory of the Earth* by James Hutton, Edinburgh, 1795.

WERNER: THE DOCTRINE OF NEPTUNISM

In retrospect, Hutton's theory of the Earth seems remarkably attuned to the then-growing scientific philosophy that the universe is rational and that everything in it is subject to unalterable natural law. The late eighteenth century was, after all, the height of the "Age of Reason." Catastrophic doctrines, however, had evolved successfully into a compromise between the Biblical story of creation and the accumulating observations of science. The prevailing viewpoint in Hutton's time held that all rocks were deposits of a primeval ocean that at one time covered the entire Earth. A definite sequence of rock types precipitated in the ocean's turbulent depths. When the water receded (where it went was never satisfactorily explained), all rocks in their present configurations and all features of the present landscape, including its deep valleys, were left behind.

On the face of it, this "Neptunist" scheme seems to us extraordinarily vulnerable to scientific observations. Limestones, granites, basalts, other igneous rocks, and metamorphic rocks as well were equally regarded as marine precipitates. However, two things combined to make this catastrophic scheme of Earth history the accepted view of its time: (1) The vast primitive sea strongly resembled the Biblical flood and thus had theological appeal, and (2) it was championed in lectures beginning in 1775 by one of the most persuasive and influential teachers of eighteenth-century Europe, Abraham Werner of Freiberg, Saxony. As a result, the catastrophic point of view held sway over Hutton's uniformitarian position until well into the nineteenth century.

Werner's Neptunist philosophy is particularly interesting to us because remnants of it still haunt stratigraphic interpretations today. A cornerstone of Neptunism was that the age of rocks everywhere can be told from their composition. Werner divided the rocks of the Earth's crust into distinct "series." The first thing deposited from the cloudy primeval ocean was granite, followed closely by a thick sequence of gneisses, schists, and other crystalline rocks. The second "series" included slate, schist, graywacke, and some limestone. Both series, subdivided into "Universal Formations," were considered to have enveloped the entire global surface at one time. The third "series," consisting of sandstone, limestone, salt, gypsum, coal, and basalt, and the fourth, consisting chiefly of sand, clay, and gravel, were deposited as the ocean waters subsided below the level of the highest mountain tops. Hence, the third and fourth "series" consisted of formations partly derived from the disintegration of older strata and partly from oceanic precipitation. These "Partial Formations" lacked total continuity because of interruptions by locally emergent mountain tops, but each of them still had worldwide distribution and a definite position in the overall sequence. Where exposed, each occurred at its own horizon and represented its own unique time of origin.

In the field, Werner's principles quickly ran into difficulties. By the 1820's geologists had come to realize that in different parts of Europe local stratigraphic sequences differ considerably. In one area granite might be overlain by a thick limestone. In another area the same granite might be overlain by a unit of alternating sandstone and shale beds. In still a third area the same sandstone and shale unit might rest on a sequence of massive dolomite instead of granite. How could these sequences differ so if the primeval ocean simultaneously produced the same rock types everywhere? This discovery undermined Werner's whole scheme, which clearly lacked predictive capability.

A second Wernerian idea that failed the test of field scrutiny was the supposed sedimentary origin of basalt. A dark igneous rock, basalt became the focus of the dispute between Werner and followers of Hutton, who held that it always crystallized from molten material that came from within the Earth. Close studies in various areas of Europe convinced one geologist after another that basalt was truly igneous in origin. A look at the magnificent array of extinct volcanoes in the Auvergne area of central France was sufficient to change the minds of even the most loyal Neptunists. There, perfect craters are the obvious source of lava streams, which flowed down nearby valleys before they cooled and froze into columnar-jointed basalts. Many geologists made the pilgrimage to the Auvergne to resolve the basalt controversy for themselves, and virtually all sided with Hutton's point of view. Subsequently the association of volcanoes and basaltic lava flows has been documented in many regions (see Fig. 1-2).

FIG. 1-2 The association of basalt flows with young volcanoes like the one shown here clearly demonstrated the igneous origin of basalt and greatly undermined the Neptunist doctrine that all rocks are sedimentary.

FIG. 1-3 James Hutton showed that Salisbury Craig in Edinburgh is a sill, and he thereby proved the reality of igneous intrusives into the Earth's crust.

However, not all basalt forms at the Earth's surface, and Hutton correctly inferred that some basalt bodies had crystallized at depth. Salisbury Craig, a prominent Edinburgh landmark, owes its relief to a thick sheet of resistant basalt. Hutton showed from the baked contacts below and above, and from places where the basalt actually invades underlying and overlying beds, that the thick basalt body was not merely a flow that had formed in sequence, but that it was *intruded* as hot magma into the surrounding sedimentary rocks long after they were deposited (see Fig. 1-3). There was simply no room in Werner's scheme for intrusive igneous bodies. The discovery of igneous intrusives thus helped to discredit Werner's simplistic layer cake still further, and it simultaneously provided new insight into the Earth's behavior.

The failure of many of Werner's ideas to stand up in the early nineteenth century cast suspicion on his entire philosophy. In the mid-1820's, however, Werner's general scheme was still taught in most centers of learning because no other widely known philosophical system was a serious contender for its replacement. Referring to these years, Adam Sedgwick, one of geology's great pioneers, recalled that, "At the time I had not quite learned to shake off the Wernerian nonsense I had been taught." Hutton's ideas of uniformitarianism, although they had been very capably popularized by John Playfair in 1802, had simply failed to capture the imagination of that generation. Thus, Neptunism's decline did not necessarily promote the decline of other cataclysmic philosophies wherein the Earth's surface was molded by forces that were seemingly more powerful than those presently at work and were thus, by implication, supernatural.

CHARLES LYELL: THE ACCEPTANCE OF UNIFORMITARIANISM

Finally, in Charles Lyell (1797–1875), Hutton's concept of gradual change through existing physical causes got an effective champion. In his *Principles of Geology* of 1830 Lyell marshaled, with lucidity and clarity, all the observations he could collect in support of the doctrine that the present is the key to the past. Almost singlehandedly, Lyell established uniformitarianism, at the expense of catastrophism, as the accepted philosophy for interpreting the history of the Earth. In so doing, he founded modern historical geology and he reintroduced, with profound impact, the concept of unlimited time. Geological problems now could be solved by reference to natural laws still active and available for study in the real world about us instead of by reference to former, shadowy, mythical, or supernatural events. About the stifling nature of the catastrophic philosophy he sought to replace, Lyell commented, "Never was there a dogma more calculated to foster indolence, and to blunt the keen edge of curiosity, than this assumption of the discordance between the former and the existing causes of change."

Lyell's wide influence prepared the ground for succeeding accomplishments of the nineteenth century, including those of Charles Darwin, whose ideas on the gradual development of living things could not have flourished without the intellectual framework of vast time. Hence, the uniformitarian doctrine was eminently successful in nourishing scientific progress. In retrospect, however, it appears that the pendulum swung a bit too far. Not only did Lyell strictly reject any process that could not be shown to accord with constant and presently verifiable laws of nature (this is proper scientific procedure in general), but he would not even entertain the thought that rates of change, or the *relative importance* of geological agents, ever differed from what they have been within human experience. In short, strict uniformitarianism possessed its own rigid and stifling aspect, brought on by allowing, for all the geologic past, only the present rates of natural processes.

UNIFORMITARIANISM TODAY

Very properly Lyell, and Hutton as well, allowed their interpretations to rest only on those objective features that can actually be observed in the rocks. Hutton saw there "no vestige of a beginning,—no prospect of an end," and Lyell confessed much later that "The same conclusion seems to me to hold true." Today, however, we observe more subtle and varied aspects of the rocks and we do see—or we think we see—a beginning. This development has put a new slant on uniformitarianism. The original concept of Lyell and Hutton envisaged a world machine endlessly repeating its cycles over and over again—a machine that was, for all practical purposes, eternal. Today we envisage an evolving planet; random local configurations may repeat themselves from time to time, but the total combination of circumstances is never quite the same twice. We think not in terms of perpetual motion, but in terms of a system that was wound up in the beginning—about $4\frac{1}{2}$ billion years ago—and has subsequently been running down.

"The present is the key to the past." This popular summary of uniformitarianism encompasses both meanings of Lyell. On one hand, it refers to the rigid concept that the rates of activity and relative importance of agents as they presently exist were always uniform in the geologic past—a descriptive theory that has not withstood the test of new data. On the other hand, it refers to the idea of the permanency of natural laws—a much firmer basis from which to proceed in all geologic investigations.

Certainly the "key to the past" statement cannot apply literally to all things. The record of life, for example, shows us a long succession of different species in the rock record, each descended from some more primitive ancestor, so that each geologic age has had its unique combination of existing species. Moreover, we recognize that rates of evolution and extinction of species have varied through geologic time. The statement can and does apply, however, to the physical and biological *laws* that govern the changes. We now think that at an early time the Earth's atmosphere lacked oxygen. Although this view may constitute uniformitarian heresy in Lyell's strict sense, we can, knowing the *natural laws* that govern the behavior of matter, predict the consequences of an oxygenless atmosphere on rocks and water and life. The point, then, is that we no longer demand of our Earth model that each of today's particular rates and special processes has prevailed unwaveringly through time past. What we do feel confident of is natural law, and this is what the old term "uniformitarianism" has come to mean. Actually, the use of the word today may suggest that geology has a unique guiding principle all its own. But it does not, and for this reason the term should perhaps be dropped. The spatial and temporal invariance of natural laws is the basic assumption of all science.

The idea of geologic activity through the influence of existing physical laws carries with it the necessary condition of an immense length of time. This concept has contributed significantly to the philosophical attitudes of modern civilization, for it provides the setting in which to perceive the Earth, and everything on it, in a state of change. To the geologist, a landscape constitutes a dynamic system whose tranquil appearance on a quiet day is like a single frame from a motion picture film. And a sequence of sedimentary rocks is a prepared historical document that has been temporarily laid aside but is destined in time to rejoin the great geologic cycle and ultimately to be reconstituted into something new.

The late nineteenth century saw the first serious efforts to estimate geologic ages quantitatively. Prior to the time of Lyell, few people wondered about the age of geologic events or the age of the Earth itself, and most of these depended entirely on philosophical speculations. Lyell's arguments beginning in the 1830's not only won the day for the uniformitarian philosophy, but generated wide interest in just how much time geologic history might actually represent. Then, following Darwin's work on evolution in 1859, the question of the magnitude of geologic time became an intensely debated focal point of the evolution controversy that lasted into the twentieth century.

ORGANIC EVOLUTION:
THE CRUCIAL REQUIREMENT OF TIME

Charles Darwin (1809–1882) did not invent the idea of organic evolution. The notion had been entertained for generations and vigorously espoused by such capable scientists as the French zoologist Jean Baptiste de Lamarck (1744–1829), a pioneer of invertebrate paleontology, and by Erasmus Darwin (1744–1802), grandfather of Charles. Until Charles Darwin's time, however, the idea had never had wide currency because the earlier workers lacked important data as well as the Huttonian concept of geologic time that is vital for the evolutionary argument. Finally, building on the uniformitarian philosophy popularized by Lyell beginning in the 1830's, Darwin constructed a singularly rational and overwhelmingly convincing argument for the origin of diverse species of organisms that populate the world. In his work, *The Origin of Species*, first published in 1859, Darwin successfully proved the existence of organic evolution to scientists and nonscientists alike, and geologic thought, indeed *all* philosophical thought, has never since been the same.

Darwin wove together a multitude of ideas from widely diverse and seemingly unrelated sources. These included observations on structural relationships among living animals, ecological adaptation of animals, individual variation within species, the principles of populations, results of the selective breeding of domestic animals, an understanding of what the fossil record means, and most important of all, the full appreciation of the tremendous length of geologic time necessary for geologic changes through processes like those acting today. Obviously geologic time was available for biological processes as well as physical ones.

Briefly, Darwin's argument can be summarized as follows:

1. Populations of animals and plants produce progeny at such a rate that were they all to survive they would increase spectacularly year after year. Long before, in 1789, Malthus had illustrated this tendency with mathematical precision.

2. The spectacular progressive increases do not, in fact, occur. Although most populations fluctuate year by year, they remain essentially constant over the long term.

3. There must be a very real struggle for existence in nature. (This idea was not new with Darwin either, but had been elaborated by Lyell and by Malthus as well.) Each individual must compete for food and must cope successfully with every facet of the environment, such as climatic extremes, diseases, and predators, in order to live to produce progeny.

4. Each individual differs from virtually all others in its species. Variation had been noticed long before—for example, by Thomas Browne, who wrote about organic diversity in 1635. By Darwin's time striking variation in domestic animals had already been produced by selective breeding. Surely the species in nature had similar potential for modification. Wouldn't some modifications in nature be better adjusted to a given environment than others?

5. Here Darwin made a break with all previous suggestions on the subject. Instead of postulating that modifications are induced by the environment and are then passed on from generation to generation, he suggested that new characters arise from within an organism *entirely by chance*, but that they do not necessarily have adaptive significance or survival value; in fact, many are downright lethal.

6. Some of the new characters are successful in coping with the environment and may even allow the organism to push beyond previous environmental barriers. Others will be unsuccessful, and individuals with these modifications will simply not survive to pass them along. This process Darwin termed *natural selection.*

The very randomness of the natural selection of very small, fortuitous variations obviously requires enormous amounts of time. Darwin felt no constraints whatsoever in this regard, for he assumed the almost limitless time of Hutton and Lyell. Darwin indicated the magnitude of time he had in mind when in the first edition of the *Origin* he estimated that 300 million years had elapsed since the last part of the Mesozoic Era. Although that estimate is now believed to be too high, the order of magnitude is correct.

Darwin took uniformitarian ideas of gradual change through existing causes and effectively extended them into the realm of life. As he explained it, life evolves continually and is always being gradually modified through natural selection. He argued that evolution is a slow process and that we see something akin to it at present only when we consciously select strains of domestic animals for particular characters. Evolution explains why diverse animals show striking similar arrangements of bones and organs; it explains how certain plants and animals can be almost unbelievably adapted to extremely specialized environments; and it explains the underlying causes responsible for the succession of different faunas and floras observed in sedimentary strata.

When questioned about the paucity of stratigraphic successions, which actually yield fossils showing all the gradual changes that are supposed to occur, Darwin emphasized the incompleteness of the stratigraphic record, and argued that probably more of geologic time is represented by breaks in the record or by barren strata than by fossiliferous ones. Today we would strongly concur.

Darwin was able to refute numerous other objections, but there were two that he was unable to cope with because he lacked critical data. The first, put forth by Fleming Jenkin in 1867, concerned heredity. Jenkin simply suggested that a favorable characteristic appearing spontaneously in an individual would be progressively swamped out and ultimately obliterated in any population group in which it occurred. When the favored individual mated with a normal one, the new characteristic would be diminished in the offspring, and when these mated with normal individuals, it would be diminished further still. In succeeding generations it would not long survive. In Darwin's time this was a difficult argument to counter. One could postulate that the new characteristic, in order to survive, must arise not in just one member, but in all members of a population simultaneously. However, this suggestion abandons the idea of the natural selection of a favorable trait, and replaces it with an abrupt total change that is uncomfortably akin to a theory of successive creations.

We now know that the answer to Jenkin's argument lies in Gregor Mendel's discovery that characters are passed through generations by units of inheritance which do not blend or mix with other units, and which can never be swamped out so long as the individual has offspring. Although Mendel reported these discoveries in 1865, Darwin died in 1882 without learning of them, and they did not become generally known until 1900.

The second objection that Darwin couldn't answer involved the very low estimates of geologic time expressed with elaborate mathematical detail by a contemporary, Lord Kelvin, who was perhaps the outstanding physicist of the nineteenth century. It was clear from the outset that the success of Darwin's synthesis depended on the availability of a tremendous amount of geologic time. As soon as the *Origin* appeared, the scope of geologic time quickly became a subject of intense interest. Friends and foes alike devoted much energy and thought to the problem of just how geologic time might be reckoned in quantitative terms. In *years*, just how much time is there? The question is a big one and the world desperately wanted an answer.

To determine the quantity of time past clearly requires a time-governed, irreversible process whose rate is known. Today we think first of radioactive decay, but in the mid-nineteenth century radioactivity was as yet undiscovered. The evolution of life itself is time-governed and irreversible, but *rates* are not constant among groups. Nevertheless, Lyell tried this approach in 1867. Lyell guessed that about 20 million years was necessary for a complete change of molluscan species, and that there were 12 such intervals since the beginning of the Ordovician Period, giving 240 million years as a rough estimate of the elapsed time. Although this was not a bad guess, it was totally unverifiable. Other kinds of approaches were numerous, but attention mainly centered about three: (1) rates of rock weathering and the corresponding increase in salinity of the oceans with time; (2) rates of accumulation of sedimentary rock strata through time; and (3) Kelvin's arguments on the rate of cooling of the Earth through time and the age of the Sun.

ESTIMATES OF GEOLOGIC TIME BASED ON SALINITY

As early as 1715 the English astronomer Edmund Halley suggested that the age of the Earth could be calculated from a study of the saltiness of the ocean. The actual plan was simply to determine the salt content of the sea with great precision and then to repeat the determination a decade later; from the expected increase in salinity one could then calculate the time required, beginning with fresh water, to reach present salinity. If this experiment was ever tried, no increase in salinity was detected. Halley clearly had no inkling of the magnitude of geologic time in the pre-Huttonian era when he wrote.

Certain workers in the late nineteenth century, reconsidering the salinity method, estimated from chemical analysis of river water the amount of sodium added to the sea each year by all the rivers of the world. Knowing the approximate volume of water in today's oceans, they estimated the length of time necessary to achieve its present salinity, assuming that the original ocean waters were fresh and that the present rate of sodium contribution by rivers was the mean for all geologic time. After making various corrections to allow for the sodium in river water that derives from oceanic salt blown inland or that is recycled from ancient marine sedimentary rock now exposed to weathering, John Joly, in 1899, concluded that about 90 million years had elapsed since water first condensed on the Earth. Joly's estimate for the age of the ocean is, we now know, far too low, chiefly because he underestimated the amount of exchange of sodium that actually takes place between sea water and rocks of the Earth's crust. We now believe that salinity of the ocean is not increasing at all, but that it represents an equilibrium condition. Sodium liberated from rocks during weathering simply does not accumulate indefinitely in the sea but is recycled, as Fig. 1-4 shows, by (1) vaporizing during evaporation and being blown inland, or (2) becoming part of the marine sediment deposited on the continents where uplift will later expose it to erosion, or (3) becoming part of marine sediment that is later reconstituted into crystalline rocks.

Another major source of error is probably the assumption that the sodium content of rivers today is representative for geologic time. Judging by what we can infer of the distribution of ancient lands and seas, continents today probably stand unusually high above sea level. If this is true, then weathering rates are probably greater than has been the average, and it follows that the present sodium content of rivers is probably greater also.

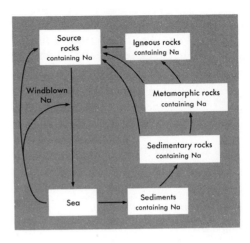

FIG. 1-4 The sodium cycle in Earth history, showing that the sea is merely one link in the cycle and not a final resting place for all sodium.

ESTIMATES BASED ON RATES OF DEPOSITION

Everyone who has studied sedimentary rocks realizes that a thick bed of sandstone may accumulate in a day; a thin bed of shale overlying it may require 100 years, and the bedding plane between them may represent more time than both put together. For a given thickness of sedimentary strata, an average rate of sedimentation exists. If the depositional environment does not change significantly and if deposition is not interrupted by episodes of erosion, then thickness of strata must be approximately proportional to elapsed time. Geologists working near the end of the nineteenth century thought that if they could establish the rate of deposition in modern sedimentary environments, they could estimate the time represented by analogous ancient rock units. They assumed further that if they could ascertain the total thickness of sedimentary rock that had ever been deposited in the geologic past, they could arrive at an estimate of total elapsed geologic time.

Troubles immediately arose because at most localities different rock types succeed each other randomly, and breaks in the stratigraphic record abound. Most workers attempted adjustments for different rock types by allowing limestone one depositional rate, shale a greater rate, and sandstone a greater rate still, and they attempted to avoid stratigraphic breaks by considering just the maximum known thickness of rocks of each age. Another difficulty was that new sections continued to be discovered that exceeded previously known thicknesses. The greatest difficulty of all, however, lay in estimating rates of sedimentation to apply to these thicknesses. Today a given kind of sediment—shale, for example—is deposited at vastly different rates at different places, and even if we could determine some sort of weighted average for the present, this would not necessarily apply to times past. In addition, rates measured for sediments require an uncertain correction for compaction before they can be applied to rocks. Furthermore, most of the geologists concerned with the problem felt that sedimentation rates in the Paleozoic Era (see the geologic time scale on the last page of the book) had been slower than those in the Mesozoic Era and that both were slower than those of the Cenozoic Era. To compensate, for different eras they used different rates, which were entirely a matter of personal judgment.

Estimates for the duration of geologic time varied widely, as Table 1-1 shows. Around 1900, however, most of them were under 100 million years as

Table 1-1   Estimates of the Age of the Earth*

Date	Author	Maximum Thickness (feet)	Rate of Deposit (years for 1 foot)	Time (millions of years)
1860	Phillips	72,000	1 332	96
1869	Huxley	100,000	1,000	100
1871	Haughton	177,200	8,616	1,526
1878	Haughton	177,200	?	200
1883	Winchell	—	—	3
1889	Croll	12,000[1]	6,000[2]	72
1890	de Lapparent	150,000	600	90
1892	Wallace	177,200	158	28
1892	Geikie	100,000	730–6,800	73–680
1893	McGee	264,000	6,000	1,584
1893	Upham	264,000	316	100
1893	Walcott	—	—	45–70
1893	Reade	31,680[1]	3,000[2]	95
1895	Sollas	164,000	100	17
1897	Sederholm	—	—	35–40
1899	Geikie	—	—	100
1900	Sollas	265,000	100	26.5
1908	Joly	265,000	300	80
1909	Sollas	335,000	100	80

[1] Spread evenly over the land areas.
[2] Rate of denudation.
* Based on estimates of maximum thicknesses of sedimentary rocks.
    After Arthur Holmes, 1913.

a result of conscious or unconscious effort to accord with Lord Kelvin's estimate of geologic time based on the age of the Sun and thermal history of the Earth. This order of magnitude, which we now know is far too low, seemed nevertheless to provide enough time for the accumulations of the known maximum total sedimentary rock thicknesses.

KELVIN'S ESTIMATES

During the period of great interest in the duration of geologic time that followed the appearance of Darwin's *Origin of Species*, Kelvin's estimates on the age of the Sun and rate of heat loss from the Earth were by far the most influential. They were also among the very lowest. Because they were based on precise physical measurements that demanded few assumptions, they seemed irrefutable, and were accepted widely, if reluctantly, by most geologists. However, Darwin and his growing following of paleontologists and evolutionary biologists could not readily accept the paltry time span that Kelvin allowed, because their theories required time of a far greater order of magnitude. Their opponents were well aware of this also. Kelvin's drastic curtailment of geologic time amounted to a flat renunciation of organic evolution through natural selection.

Heat Loss from the Earth

Temperatures in deep mines of many areas had shown a substantial and fairly uniform temperature increase with depth. This thermal gradient indicates that heat is flowing from the hot interior to the cool outer portion of the crust where it escapes. The heat loss can be measured, and Kelvin reasoned that, if in losing heat the Earth is becoming progressively cooler, then in times past it must have been warmer. The further back in time, the warmer it must have been. Kelvin regarded this phenomenon as dissipation of heat from an originally molten condition, and he showed from the present rate of heat flow that it surely was not very long ago, in terms of geologic time, that the Earth was molten. This apparent time of crystallization of the Earth's solid crust established the maximum possible age of life as we know it. Lack of details on melting points of rocks and thermal conductivity at high temperatures and pressures prohibited really precise estimates of the time of crystallization, but the order of magnitude was low. The talk was always of less than 100 million years, and regarding the American Clarence King's estimate in 1897 of 24 million years, Kelvin said, "I am not led to differ much."

These assertions had ominous overtones for evolution, not only because of the short length of geologic time they allowed, but also because, old or young, the Earth's crust had been hot in the not-too-distant past. It could not have maintained the long-term stability that Darwin deemed necessary for the gradual evolution of the kind of life that inhabits our planet.

Age of the Sun's Heat

Scientists of the late nineteenth century realized fully the fantastic energy output of the Sun, and in that period, before they had any inkling of radioactivity, the source of the Sun's energy loomed as a substantial problem. A popular theory of the day held that the energy source lay in a gravitational contraction of the Sun's mass. This process would provide a relatively longer-lived Sun than if it were simply burning, but still one that would spend itself rapidly. Regardless of the precise mechanism of solar energy, Kelvin reasoned, the great dissipation of energy forbade the idea that the Sun's heat was inexhaustible. Surely the persistent loss of so much heat energy by radiation must gradually lower its temperature. He concluded on the best grounds then available that the Sun has probably illuminated the Earth for only a few tens of millions of years.

Only a million years ago, according to Kelvin, the Sun was providing the Earth with much more energy than it is now, and in a few million years it will be providing the Earth with much less. We know that 10 percent more light and heat would destroy us, and so would 10 percent less. Approximately uniform solar energy is thus essential to the continuity of life on Earth, and this is what Kelvin, by way of the Sun, was denying the evolutionists. This argument, like that concerning the Earth's thermal history, portrayed the Earth as a substantially different place to live in times past than it is now. In 1897 Kelvin summarized his ideas on the subject for the last time and concluded that the Earth had probably been habitable for between 20 and 40 million years.

Darwin could only admit that Kelvin's data constituted a formidable objection to natural selection. In the confused intellectual climate in which Darwin penned later additions of the *Origin*, he retreated from his original firm position on natural selection. He removed concrete references to enormous time spans and he attempted to compromise his previous extremely slow evolution rates. In short, his whole theoretical structure had become shaky owing to attempted adjustments to the arguments of Jenkin and Kelvin. Already most geologists had simply adjusted their ideas of the duration of geologic events to accommodate Kelvin's foreshortened time framework. Paleontologists and evolutionary biologists who refused to go along with Kelvin's estimates could offer in reply only qualitative arguments that, in the face of Kelvin's quantitative arguments, were largely ignored. In the long run, however, their seemingly vague hunches prevailed. We have already seen how Jenkin's arguments were ultimately destroyed by Mendelian genetics. Kelvin's quantitative proofs were to undergo a similar fate in the light of radioactivity.

DISCOVERY OF RADIOACTIVITY

In 1896 a French physicist, Henri Becquerel, discovered that uranium emits mysterious rays that can actually activate photographic plates in total darkness. Becquerel termed this totally unanticipated property of uranium *radioactivity*. Within a few short years the significance of radioactive energy in geologic processes became apparent. From the Paris laboratory of Pierre and Marie Curie in 1903 came the discovery that a sample of radium always maintains a temperature greater than that of its surroundings. Three years later an English physicist,

R. J. Strutt, estimated the quantity of heat that is continuously generated by radioactive minerals in the Earth's crust and showed that this easily accounted for the flow of heat from the surface. With this discovery, the heat escaping from the Earth's surface ceased to signify impending cooler times ahead and markedly warmer times past, because the heat no longer needed to be considered residual. Instead, it is always being produced within the Earth by radioactive processes: thus, heat loss from the Earth's surface has been about the same for a very long time.

In addition, radioactive energy seemed to be the key to the mystery of the Sun's heat. Although the precise mechanism of nuclear fusion was not immediately understood, it was apparent, in light of the newly discovered energy source within the atom, that Kelvin's view of the Sun as some sort of dwindling coal pile was no longer necessary. Radioactivity replaced the youthful Sun of Kelvin with one that is vastly longer-lived and that, in the eyes of astronomers, has maintained a nearly constant energy output since it formed about 4.6 billion years ago.

Within ten years after Becquerel's discovery of radioactivity in uranium, it had been discovered in thorium, rubidium, and potassium as well. Ernest Rutherford and Frederick Soddy in 1902 postulated that radioactive elements actually change into other elements during emission of radioactive rays, one of the products of uranium decay being helium from alpha particles. Then in 1906 Rutherford, in England, made the first attempt to measure the ages of minerals from their helium-uranium ratio. Tests on the leakage of helium from the minerals indicated to early workers that at best only minimum ages could be obtained. Even so these initial measurements gave correct orders of magnitude for geologic time.

In 1905 B. B. Boltwood, an American chemist, concluded that, besides helium, lead was a stable end product of the decay of uranium. Boltwood went on to show in 1907 that in unaltered minerals of the same age, the lead-uranium ratio is constant, but among those of different ages, the lead-uranium ratio is substantially different: the older the mineral, the greater the ratio. Using the rough early estimates of decay rates, Boltwood calculated tentative radiometric ages for several mineral assemblages. Some of his results are shown in Table 1-2. Considering the sketchy knowledge concerning decay rates and the primitive analytical techniques of his day, Boltwood's radiometric dates are commendably close to those we obtain for the same rocks today.

In 1911 Arthur Holmes, then a student working in Strutt's lab in England, wrote his first paper on radioactive dating, in which he set down many of the principles that were to guide radiometric work during the following decades. Holmes summarized all radiometric age data then available, verified Boltwood's observation that lead was the end product of uranium decay, outlined the problem of sample contamination, presented evidence that decay rates are constant under any circumstances, and pointed up the potential of radiometric

Table 1-2 Boltwood's First Radiometric Dates

Geologic Period	Lead/Uranium	Millions of Years
CARBONIFEROUS	0.041	340
DEVONIAN	0.045	370
PRECARBONIFEROUS	0.050	410
SILURIAN OR ORDOVICIAN	0.053	430
PRECAMBRIAN		
SWEDEN	0.125	1,025
	0.155	1,270
UNITED STATES	0.160	1,310
	0.175	1,435
CEYLON	0.20	1,640

After Arthur Holmes, 1911.

age determinations in unraveling Precambrian history. In addition, he clearly foresaw the day when the periods of the geologic time scale might be graduated with accurate radiometric dates.

In addition to refuting, once and for all, Kelvin's arguments demanding a young Earth and Sun, radioactivity thus provided the ultimate tool for measuring geologic time. It was clear from even the earliest efforts that the Earth's antiquity was far greater than prevailing scientific opinion at the turn of the century would have deemed possible. The ages provided by the new radiometric tools would surely have pleased Darwin, because they were of the scope which he had argued arduously for, on altogether different grounds, a half century before.

THE MAGNITUDE OF GEOLOGIC TIME

Even today the actual quantity of elapsed geologic time, because it is so unimaginably great, means little without some basis for comparison. To this end numerous schemes have been invented in which key geologic events are apportioned their proper places in every day units of length or time in order to make geologic time somewhat more comprehensible.

Compress, for example, the entire 4.6 billion years of geologic time into a single year. On that scale, the oldest rocks we know date from about mid-March. Living things first appeared in the sea in May. Land plants and animals emerged in late November and the widespread swamps that formed the Pennsylvanian coal deposits flourished for about four days in early December. Dinosaurs became dominant in mid-December, but disappeared on the 26th, at about the time the Rocky Mountains were first uplifted. Manlike creatures appeared sometime during the evening of December 31st, and the most recent continental ice sheets began to recede from the Great Lakes area and from northern Europe about 1 minute and 15 seconds before midnight on the 31st. Rome ruled the Western world for 5 seconds from 11:59:45 to 11:59:50. Columbus discovered America 3 seconds before midnight, and the science of geology was born with the writings of James Hutton just slightly more than one second before the end of our eventful year of years.

Those concerned with the total age of the Earth commonly consider *the beginning* as the time when the Earth achieved its present mass. Probably this was essentially the same point at which the Earth's solid crust first formed, but we have no rocks that date from this early time. Indeed, evidence now available suggests that no rocks have survived from the first several hundred million years of Earth history. Prior to the beginning, cosmic processes were giving rise to matter as we now know it and to our solar system. This interval we put down to cosmic time. It is the time since the beginning of the Earth that properly constitutes *geologic time*.

two

the rock record

The rocks exposed at the Earth's surface or within reach of our drills constitute our only record of Earth history. They are materials rearranged by natural processes during bygone ages, and they testify by their physical attributes and their fossils to the characteristics of the environmental settings in which they formed. Each rock has its own history, its own story to tell. It is important for our understanding that we learn to read them, for in the geologic past there was no one around to observe the events and to write down what they saw.

Rocks may be grouped into three main categories based on their mode of origin: igneous, sedimentary, and metamorphic. Igneous rocks crystallize from molten material; those that form within the Earth are called intrusive or plutonic, and those that form at the Earth's surface are extrusive or volcanic. Sedimentary rocks originate from the compaction and cementation of material that has been weathered from pre-existing rocks and transported to a place of rest by water, wind, or ice. Metamorphic rocks are former sedimentary or igneous rocks that have been recrystallized by heat, mineralized solutions, and usually the accompanying pressure of deep burial, but without melting.

SEDIMENTARY ROCKS

Of the three kinds of rocks, sedimentary rocks provide by far the most complete record of Earth history. First of all, they constitute about 75 percent of all exposed rocks; second, they alone form at normal temperatures and pressures at the Earth's surface; finally, they are the only rocks that generally contain fossils, and fossils not only record the history of life, but they are by far the best tools we have for correlation. The singular importance of sedimentary rocks as documents of geologic time gave rise in the mid-nineteenth century to the term *stratigraphy* to apply to the study of their history.

Sedimentary rocks are stratified; this is their most conspicuous characteristic (see Fig. 2-1). Each stratum is a layer of rock that represents an individual episode of *sedimentation*, or deposition. Successive strata are separated by boundaries called *bedding planes*. Bedding planes may occur at marked changes in rock type, as in alternating beds of sandstone and shale, or at subtle changes in cementation or grain size in a sequence of uniform lithology. Individual strata, depending largely on the sedimentary environment in which they formed, may be many feet in thickness or paper thin, and they may extend laterally only a few centimeters or for hundreds of kilometers. In any event each sedimentary layer and its enclosing bedding planes indicate the passage of time. The geologic systems that provide the calendar for Earth history are all defined in areas of sedimentary rocks. Each is dated relative to the others by its position in the worldwide sequence established by correlating the rock records of many localities.

FIG. 2-1 Alternating sandstone and shale beds of varying thickness —strata of probable earliest Cenozoic age near Santa Cruz, California. (W. C. Bradley.)

Classification

A simple classification, suitable for our purposes, recognizes two main categories of sedimentary rocks, detrital and chemical (see Table 2-1). Detrital rocks are composed of particles of fragments of material derived from pre-existing rocks and they are classified by the *size of constituent particles* into conglomerate, sandstone, siltstone, and shale. Chemical rocks are usually of organic origin or they are chemical precipitates. They are classified on the basis of their *mineral composition* into limestone, dolomite, salt, gypsum, chert, and coal. The twofold classification is not entirely natural; on occasion chemical sedimentary rocks may be mechanically eroded and the resulting detritus reconstituted into new strata with the same composition. Such a so-called chemical sediment would thus be of detrital origin. Moreover, many sedimentary rocks contain both detrital and chemical components. In this case adjectives express the minor component, as in "shaly dolomite" or gypsiferous sandstone."

Individual rock bodies that are composed of one of the rock types listed in Table 2-1 are thus not all alike by any means. Many different kinds of each

Table 2-1   The Common Kinds of Sedimentary Rocks

DETRITAL ORIGIN

Rock Type	Particle Diameter
CONGLOMERATE	Pebbles—2 to 64 mm
SANDSTONE	Sand—$\frac{1}{16}$ to 2 mm
SILTSTONE	Silt—$\frac{1}{256}$ to $\frac{1}{16}$ mm
SHALE	Clay—less than $\frac{1}{256}$ mm

CHEMICAL OR BIOCHEMICAL
(LESS COMMONLY DETRITAL) ORIGIN

Rock Type	Composition
LIMESTONE	Calcite—$CaCO_3$
DOLOMITE	Dolomite—$CaMg(CO_3)_2$
SALT	Halite—$NaCl$
GYPSUM	Gypsum—$CaSO_4 \cdot 2\,H_2O$
CHERT	Silica—SiO_2
COAL	Chiefly carbon

can be distinguished through details of texture and composition. Shale, the most abundant sedimentary rock, is somewhat difficult to study in detail because the individual clay particles cannot be resolved with ordinary microscopes. Electron microscopes, X-ray techniques, and other sophisticated methods have become widely utilized, and in recent years we have learned much about clay minerals and the shales they form. Sandstones and limestones have been studied in the greatest detail, partly because they are important sources of oil and gas and partly because they are easiest to study. Let us briefly survey some of the subtypes of sandstones and limestones in order to illustrate the kinds of interpretations they permit.

Kinds of Sandstones

When we study sandstones in detail, we investigate both texture and composition. Texture encompasses the parameters of grain size, shape, and sorting. Sorting refers to the *range* of grain sizes that constitute the sandstone. If the range of sizes is small, the sandstone is well-sorted. If the range of sizes is great, the sandstone is poorly sorted. Very fine detrital material that fills the spaces among the larger grains in poorly sorted sandstone is called *matrix*. Textural details, and compositional details as well, are most easily viewed in *thin sections*, rock slabs that have been ground thin enough to transmit light. Figure 2-2 shows sketches of a well-sorted and poorly sorted sandstone.

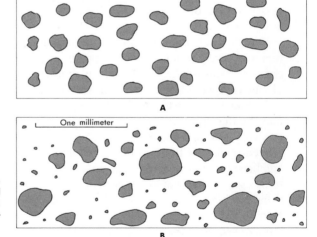

FIG. 2-2 (A) Sample of a well-sorted sediment. Most grains fall within the same size range. (B) Sample of a poorly-sorted sediment. Grains range widely in size.

The composition of a sandstone is determined primarily by the detrital material, but partly by the *cement*, which is material precipitated chemically in the spaces among the grains during lithification. All spaces between the grains of a sandstone are not typically filled; some generally remain as open pores allowing the sandstone to transmit fluids, in which case the rock is said to be permeable. If the spaces between the grains are almost completely filled with cement or with matrix, the sandstone is incapable of transmitting fluids and it is said to be impermeable, or "tight."

Sandstones formed almost entirely from quartz grains are called quartzose sandstones or *orthoquartzites*. They may contain small amounts of other very stable minerals, but they contain little clay matrix and virtually no unstable minerals. For this reason orthoquartzites are referred to as "mature." They typically represent slow sedimentation in stable regions where the sediment may be extensively winnowed and reworked before it finally comes to rest. Many orthoquartzites were derived from pre-existing sandstones, in which case they are "second cycle" sedimentary rocks as opposed to those derived directly from igneous or metamorphic terrains.

Sandstones that contain a low proportion of clay and silt matrix and a great many feldspar grains are called *arkoses*. Arkoses are derived from granitic source areas where mechanical weathering is dominant, and the combination of quartz and feldspar gives arkose a mineralogical composition similar to granite.

Sandstones that have a great deal of matrix are called *graywackes*. Graywacke grains generally consist of rock fragments, feldspar, and quartz. The heterogenous composition and poor sorting of graywackes has always been taken to mean environmental instability, that is, rapid erosion in the source area and, in particular, rapid deposition in the depositional area. Sometimes graywackes are described as having a "poured in" appearance, as a result of their apparent lack of winnowing in the depositional area. Lately, sedimentologists have been asking two questions: (1) Why do we see so few graywacke-type sediments being laid down in areas of rapid deposition today? (2) If graywackes represent really rapid sedimentation and rapid deposition, why do they so rarely contain grains of chemically unstable minerals like amphiboles, pyroxenes, and olivine that would be expected to survive such circumstances?

Some geologists now feel that the lack of unstable minerals in graywackes may be due to their later destruction by chemically active ground waters. In this view the unstable grains were initially present but have subsequently been altered to clay. If this is true, the influence of postdepositional chemical change has been underestimated in the past. The "poured in" appearance that characterizes graywacke would not be primary but secondary, the clays having resulted from the breakdown of unstable sand-size grains. This suggestion would also help to explain why we see so few graywacke deposits forming presently in modern sedimentary environments. However, moderately well-sorted sediments containing a high percentage of unstable minerals, sediments that are forming today in many places, may alter at a later time to graywacke and acquire a "poured in" appearance.

Kinds of Limestones

Among the chemical sedimentary rocks, limestones have received the most study; as a result several kinds can be distinguished based on the shape and origin of the calcium carbonate particles that make them up. Collectively

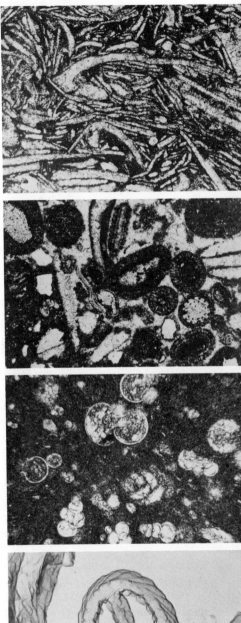

FIG. 2-3 Microscopic views showing diverse components of limestones. (A) A limestone composed of fossil shell fragments in a fine-grained matrix (magnified 17 times). (B) A limestone composed of oölites (round grains) and skeletal particles in a clear crystalline matrix (magnified 24 times). (Both from A. Ford and J. J. H. C. Houbolt, 1963.) (C) Dense, very fine-grained limestone with a few planktonic foraminifers (magnified 140 times). (D) Electron microscope photograph of limestone like that shown in (C) reveals that extremely tiny algal plates called coccoliths constitute much of the fine-grained material (magnified 18,000 times). (Courtesy W. W. Hay.)

they represent a host of environments, and about the only generality applicable to all limestones is that they form in areas where large quantities of terrigenous (land-derived) detritus are not available. Beyond this, the water may be shallow or deep, marine or nonmarine, and the immediate sedimentary agent inorganic or entirely organic. For example inorganic limestones may consist of fine-grained deposits formed in very quiet water, or of oolites, sand-sized spherical particles formed under intense agitation. Limestones of biochemical origin may consist of ultramicroscopic plates from the smallest algae or of heavy skeletons of reef-building corals. In short, the variety is great, and it results almost entirely from differing depositional environments. Thin sections of some diverse kinds of limestones are illustrated in Fig. 2-3.

Like sandstones, limestone deposits are sometimes substantially changed long after they have been deposited. Some limestones have altered partly or entirely to dolomite, due to the actual introduction of magnesium by circulating ground waters. Others have altered partly or entirely to chert. The evidence in both cases is usually quite clear, including for example, dolomitized or silicified fossil shells known to have been originally calcareous.

MAJOR PATTERNS OF SEDIMENT DISTRIBUTION

Large-scale environments on the Earth's surface, today as in the past, can be divided into those of erosion and those of sedimentation. Material for sedimentary rocks originates in the former and collects in the latter. Areas of erosion are typically areas where uplift has occurred, and areas of accumulation, are typically areas of net subsidence. Deposition of sediment may occur in the sea, on the land, or in transitional environments between the land and sea. In order for a sedimentary record to be created in one place, older rocks must be eroded in another. Figures 2-4 and 2-5 provide a present-day example. The creation of a sedimentary rock is thus really just a rearrangement of material at the Earth's surface. In being transferred from one place to another, material is shaped by the prevailing environments and as a consequence becomes a record of them.

Whether an area of sedimentation is subaerial or submarine, the sediment surface, as a general rule, is of fairly low relief. In erosional areas, relief may be great, as in mountainous regions, or small, as on plains. Sedimentary rocks exposed in areas presently being eroded testify that what was once an environment of sedimentation has been changed by uplift to an environment or erosion. Changes from environments of sedimentation to those of erosion and back again have typically occurred many times in one area during geologic history. Times of erosion or nondeposition produce significant stratigraphic disconti-nuities called *unconformities*. No area has received sediments continuously throughout geologic history, but some areas have much more complete strati-graphic records than others.

FIG. 2-4   These rugged sandstone pinnacles near the junction of the Green and Colorado Rivers in southern Utah (see map in Fig. 2-5 for location) are produced by the erosion of an ancient sedimentary record. In the process of its destruction much of it is exposed for our inspection. (From U.S. Geological Survey, 2nd Ann. Rept., Pl. XXII.)

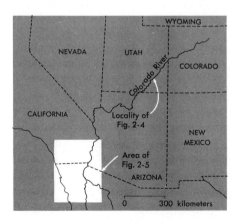

FIG. 2-5   Delta of the Colorado River. Here a vast new depositional record is being created by the material eroded from the high plateaus and mountains to the northeast and carried by the Colorado to the sea.

Geosynclines and Platforms

The thickest accumulations of sedimentary rocks occur in great belts, hundreds of miles long, called *geosynclines.* Youthful geosynclines, like the Cenozoic Gulf Coast Geosyncline, and most older ones, like the Paleozoic Cordilleran Geosyncline on the west border of the United States and the Paleozoic Appalachian Geosyncline on the east border, lie at the margins of continents. Some older ones, like the Paleozoic Uralian Geosyncline of the U.S.S.R., slice across continental interiors. From being belts of rapid subsidence where thicknesses typically of 12,000 meters or more of sediments accumulated, many geosynclines have been transformed into great mountain chains. Thus geosynclines are not permanent: they are formed; they subside as sediments accumulate, usually for a few geologic periods; and then they expire, usually by intense crustal deformation. While they exist, geosynclines receive thicknesses of sediments several times as great as do the stable areas between them, and thus unconformities in geosynclines are fewer and of less duration than in stable regions.

Stable regions that typify interior areas of continents are called *platforms.* Subsidence on the platform areas has, for the most part, been comparatively small and has permitted total accumulation of only a few thousand meters of sedimentary rocks. Unconformities are numerous and the geologic record is far less complete than in geosynclines. Some portions of platform areas have persistently subsided less and have been emergent more often than their surroundings. As a consequence the total accumulation of sedimentary rocks on these *domes* or *arches* is thinner and unconformities are even more numerous than in surrounding areas. Certain other portions of platforms have persistently subsided more than average and hence have received a thicker and more complete sedimentary record than surrounding areas. Those that are distinct and roughly circular are termed *basins.*

SEDIMENTARY ENVIRONMENTS

Ancient environments can be reconstructed only for areas in which sedimentation occurred. There has to be a rock record. The Earth's vast regions of uplands and mountains today contain a fascinating array of biotic and physical environments, but these are not generally preserved in the Earth's historical record. Although uplands and mountainous areas receive temporary deposits that may provide an informative record of recent geologic history, these sediments are not destined to become buried and lithified into rock. Over the long term the landscapes in which they exist undergo net erosion, and temporary sedimentary deposits are removed to subsiding lowlands, commonly in or adjacent to the sea, where a significant historical record is accumulating. The Earth's historical record is thus biased in that it chiefly contains information about environments of sedimentation, and relatively little information about the source regions of the sediment.

Sediments are responsive to environmental variables. The exact characteristics of a given sedimentary rock depend on (1) the kind of sediment from which it formed and (2) how the sediment was arranged and modified in the depositional area. The kind of sediment that is supplied to a depositional area depends on (1) the composition of the parent material in the source area, (2) the climate and relief in the source area, and (3) the mechanism of transportation. Just how the sediment is arranged and modified in the depositional area depends on existing environmental factors, including climate, wave and current energy, rate of subsidence, pH, salinity, water temperature, and rate of detrital influx. In depositional areas where little or no detritus is available from an outside source, chemical sediments alone may form. Biologic activity is also an important environmental factor for detrital as well as chemical sediments, both during sedimentation and after. For example, burrowing organisms may greatly modify sediments after deposition but prior to deep burial (see Fig. 2-6). In some sedimentary rocks, in fact, burrowing activity has totally destroyed the original stratification.

FIG. 2-6 Siltstone and shale laminae (layers) disturbed by burrowing organisms prior to lithification—Cretaceous Skull Creek Shale, northern Colorado.

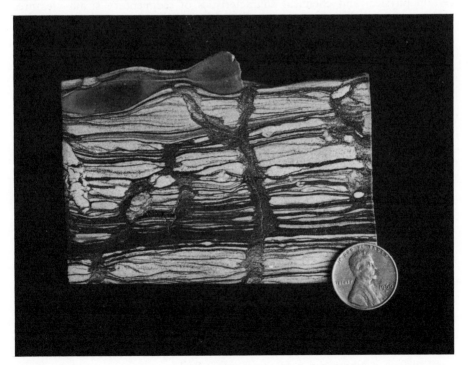

Characteristics of sedimentary rocks are thus influenced not only by actual sedimentary environments but also by the postdepositional environments in which lithification and diagenesis occur. *Lithification* refers to all processes that convert newly deposited sediment into rock, including compaction of the sedimentary particles and their cementation by precipitated mineral matter. *Diagenesis* refers to chemical reactions that take place between minerals and fluids within the sediment or sedimentary rock. It includes all changes up to, but excluding, metamorphism. Diagenesis overlaps with lithification, and both depend at first on factors within the depositional environment.

In summary, sedimentary rocks are documents of environments as well as documents of time. It is not possible, nor is it particularly desirable, to interpret one without due regard for the other. Our chief goal is to understand how the two elements interrelate—to be able to draw, for any given moment in the geologic past, a map that shows the areal extent and lateral relationships of the then-prevailing environments. Although our focus in this book is primarily on time, it is appropriate to digress for a moment and to elaborate briefly on those environments that are imprinted in the rock record.

At any given time in the geologic past environmental factors governed the geographic distribution of plants and animals, and hence the fossil content of sedimentary rocks. Environmental factors also governed the distribution of certain physical and chemical characteristics of sedimentary strata—features such as composition, bedding, and other sedimentary structures. In some cases particular physical and chemical characteristics permit us to quantify specific environmental factors: glacial deposits signify temperatures below 0°C; salt beds precipitate from normal sea water only when evaporation raises the density to 1.21; reef corals thrive only in warm sunlit marine waters of normal salinity; coarse gravel may be transported by water currents only if they achieve velocities approaching one meter per second. The sediments formed under these and similar circumstances tell us a great deal about environmental specifics. In many other cases, however, we do not understand the influence of individual environmental factors on sediments because the factors are interdependent and because, in modern environments, their effects are difficult to observe. In order to inspect three-dimensional aspects of existing sedimentary environments, we must dig down with a shovel or coring device, activities that have their practical limits. Shovels are useless in the all-important submarine environments and cores give little information on lateral extent and variation of the beds. Moreover, even those features that can be observed may have no apparent relationship to currently operative surface processes. Even though we may know that a particular feature formed in, say, a shallow sea or a river bed, the precise circumstances that governed its origin may completely elude us. The obvious solution is to watch features actually form, but again practical problems arise. Many significant features form during times of extraordinary energy, such as hurricanes and floods. During these events, observations are difficult; the sediment surface is submerged, and commonly it is about all an observer can do to survive, to say nothing of making measurements, recording data, and taking photographs. But progress is being made. With the aid of periodic observations in sedimentary environments, and through laboratory experiments that can simulate certain natural conditions, we are learning more and more about cause and effect in physical sedimentation.

Table 2-2   Classification of Sedimentary Environments

CONTINENTAL	**Alluvial** (stream channel and floodplain deposits; alluvial fans)
	Lacustrine (lakes of humid or arid regions)
	Eolian (wind-blown deposits, chiefly in deserts)
	Swamp (in poorly drained areas not associated with the seashore)
	Glacial (includes deposits of outwash streams as well as the ice itself)
MIXED OR SHORE- RELATED	**Delta** (a composite of alluvial distributaries, marsh, bay, and shallow marine environments)
	Estuary (drowned valleys)
	Bay, Lagoon
	Marsh (poorly drained areas at sea's margin)
	Intertidal and Supratidal Flat
	Barrier Island and Beach
	Glacial-Marine (ice-rafted deposits)
SHALLOW MARINE (Neritic, depth up to 200 meters)	**Shelf Banks**
	Shelf Basin (unrestricted, restricted)
	Graded Shelf
	Carbonate Shelf and Reef
	Evaporite Basin (most are shallow, some deep marine)
DEEP MARINE (Bathyal, 200- 3,700 meters; abys- sal, deeper than 3,700 meters)	**Continental Slope and Submarine Canyon**
	Submarine Fan
	Deep Ocean Basin (pelagic, terrigenous)
	Deep Enclosed Marine Basins

For most purposes we presently use a geomorphic classification of sedimentary environments like that in Table 2-2. The geomorphic setting—for example, whether a sediment formed on a river's flood plain or on a tidal flat—is of primary interest, but we would also like to be able to infer more about specific environmental factors, such as slope of the flood plain, temperature and salinity on the tidal flat, and current velocity in both cases. We are able to add details to this classification as we learn more about relationships between individual environmental parameters and particular characteristics which they impose on sediments. Important in this regard are sedimentary structures that are noteworthy features of sedimentary rocks and helpful indicators of environments.

SEDIMENTARY STRUCTURES

Stratification

Most bedding planes that separate strata represent pauses in sedimentation. Others are caused by a change in the type of sediment being deposited. The change across bedding planes may be obvious, as in a change in composition of material, or subtle, as in differing degrees of sorting in successive sandstone beds. In many sandstone and limestone units, strata are made clear by their shaly partings. If partings are not present, limestone and sandstone units of uniform hardness may appear quite "massive"—that is, without apparent stratification. Strata less than one centimeter thick are usually called *laminae* (see Fig. 2-6) and those more than a centimeter thick are *beds* (see Fig. 2-1). Thickness of strata may be described following the scheme in Table 2-3.

Table 2-3 Classification of Strata According to Thickness

BEDS	Very thick-bedded		
		100	centimeters (about 3 feet)
	Thick-bedded		
		30	centimeters (about 1 foot)
	Medium-bedded		
		10	centimeters (about 4 inches)
	Thin-bedded		
		3	centimeters (about 1 inch)
	Very thin-bedded		
		1	centimeter (about 0.4 inch)
LAMINAE	Laminated		
		0.3	centimeter (about 0.1 inch)
	Thinly laminated		

In environments where currents sweep sediment into bars, dunes, ripples, or other irregular surfaces, bedding actually forms at an angle to the principle surface of accumulation, and the deposit is said to be *cross-bedded* or *cross-stratified*. Each individual stratum thus deposited at an angle to the attitude of the formation as a whole is a *cross-stratum* or *foreset bed*. A sequence of parallel cross-strata is called a *set*. Sets may be very tiny, like those formed by ripple lamination, or they may be composed of beds many feet thick and hundreds of feet long. Cross-bedded units may be divided into (1) those in which sets have essentially planar contacts and are thus *tabular* bodies and (2) those in which they have curved basal contacts and are thus *trough-shaped*. Between these two principle kinds of cross-bedding (see Fig. 2-7), gradations exist. On outcrops, distinguishing between tabular and trough cross-bedding commonly depends on being able to observe the structures in three dimensions. In exposures that are cut parallel to the direction of transport, both kinds may appear to have essentially parallel contacts, and both may contain cross-strata that are tangential or nontangential to the basal surface of the set.

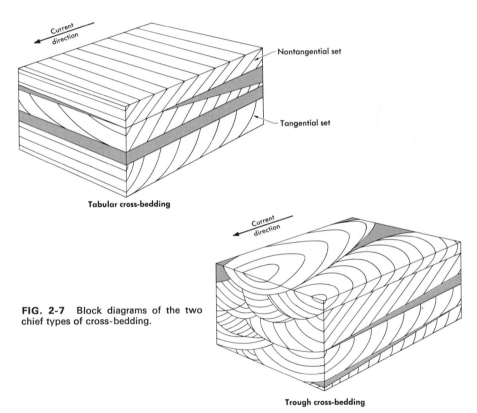

FIG. 2-7 Block diagrams of the two chief types of cross-bedding.

Structures On and Within Strata

Several kinds of structures may occur at bedding planes as a result of markings on the *bottom*, or "sole," of the upper stratum or on the *top* of the lower stratum. Markings on the base of sandstone or limestone beds are generally casts of impressions formed on the surface of an underlying shale bed when it was still soft mud. These include mainly (1) load structures, which result from irregular downsinking of sand into underlying soft mud during compaction, (2) current structures, which are due to the action of currents or to material swept along by the current on a mud surface, and (3) casts formed in animal trails or burrows. Structures on top of beds include (1) ripple marks, (2) mud cracks, (3) erosional marks formed by currents, (4) tracks and trails of animals, and (5) imprints of rain drops and bubbles.

FIG. 2-8 Some sedimentary structures of value in delineating top from bottom of beds. (A) Tangential cross-bedding. Note how the tops of cross-bedded sets are truncated. Jurassic Navajo Sandstone, Utah. (Courtesy W. C. Bradley.) (B) Oscillation ripple marks on a single bedding plane opened like pages in a book. Piece on bottom is upside down. Triassic Lykins Formation, Colorado. (C) Sole markings including load structures and current structures on the base of a sandstone bed. Pennsylvanian Minturn Formation, Colorado.

A

B

C

The most common structures that occur *within* beds are concretions and disrupted or deformed bedding. Concretions are normally spheroidal, post-depositional accumulations of mineral matter in a sedimentary rock. They range in size from small pellets to large bodies 2 or 3 meters in diameter. Disrupted bedding may be caused by slumping or sliding of soft sediment prior to lithification, or it may be due to migrating fluids or burrowing animals. Bedding may also be deformed by differential compaction about fossils or concretions.

Sedimentary structures can be a great aid to environmental interpretation. Many, however, can form in more than one environment and hence the ancient and modern must be compared with caution. Current ripple marks on strata, for example, because they are extremely common in modern shallow-water environments, were widely interpreted in the past as representing shallow-water deposits. Thanks to photography of the deep sea floor in recent years, however, rippled sediment surfaces have been observed at depths of several thousand meters, an indication that currents sweep these areas. It is now clear that current movement is all that ripple marks indicate about ancient strata. Although they tell nothing of depth, current ripples do indicate the direction the current was moving, and this is a valuable aid in reconstructing the ancient environment.

Besides their value in environmental interpretations, some structures indicate the correct order of stratigraphic succession in areas where strata are tightly folded, steeply dipping, and even metamorphosed. In fossiliferous rocks the fossil sequence alone can often provide the order of succession. In poorly fossiliferous or unfossiliferous strata, which includes all Precambrian terrains, a great many kinds of sedimentary structures may provide the best answer to "which way is up?" These include structures on and within the beds. A few of the most useful structures in discriminating top and bottom in sedimentary sequences are illustrated in Fig. 2-8.

THE LOCAL SECTION

Although most of the Earth's land surface is covered by a thin layer of soil, it is usually easy to determine the nature of the underlying bedrock, either by digging through the soil or, more commonly, by observing the rock where it is exposed in hills and stream cuts or in man-made excavations. In addition, data from surface exposures may be supplemented by information on the "subsurface" rock distribution in areas that have been drilled.

To understand the geologic history of an area one must first distinguish the individual bodies of rock that occur within it, and then one must organize these rocks into a chronologic sequence. The chronologic sequence of stratified rocks in a given area is called the local *stratigraphic section*. This is commonly illustrated as a vertical strip such as that shown in Fig. 2-9. Once established, the local stratigraphic section may be interpreted for the environments it represents.

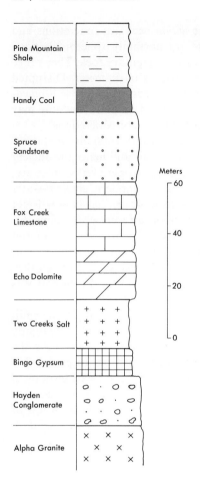

Pine Mountain
Shale

Handy Coal

Spruce
Sandstone

Fox Creek
Limestone

Echo Dolomite

Two Creeks Salt

Bingo Gypsum

Hayden
Conglomerate

Alpha Granite

FIG. 2-9 Stratigraphic column for a hypothetical area showing symbols for common rock types.

Meters
60
40
20
0

Rock Units

The stratigraphic section for a given locality is made up of rock units, bodies of distinct rock types that are distinguishable in the field.

The fundamental rock units are called *formations.* Thickness of individual formations, measured perpendicular to bedding, is usually several tens or hundreds of meters, but some are much thinner and others much thicker. Lithologic identity is the sole basis on which formations are defined. A given formation is worthy of recognition and naming if it is distinctive enough and thick enough to be mapped. The name consists of two parts: the first is a geographic locality where the rock is well-exposed; the second describes the general rock type. The St. Louis Limestone and the Prospect Mountain Quartzite are typical formation names used in mapping. For convenience two or more formations are sometimes lumped into larger rock units called *groups*. Subdivisions of formations are called *members*.

Individual formations are distinctive because each is the product of a particular depositional environment or an alternation of related depositional environments. Environments responsible for particular kinds of sediments typically shift laterally with time. In most cases a kind of rock that was produced in a given ancient sedimentary environment began deposition at different times in different places and ceased deposition at different times in different places. The boundaries of recognizable rock units, therefore, do not commonly coincide with time boundaries but usually transgress them as they are traced laterally.

Major problems in interpretation arise when mapped rock units are treated as subdivisions of geologic time (that is, as time-stratigraphic units). Yet, this was fairly common practice until around 1940, when rock units were at last generally recognized as a totally separate category of stratigraphic unit, free from any time connotation.

Geologic Maps

In studying the geology of a region, geologists normally begin by constructing a geologic map, which shows the distribution of different formations as they would look if all the soil were removed from the surface. An accurate geologic map shows exactly how the various formations are arranged. From a good map, one who has never actually visited an area can gain a good understanding of it. An indication of the utility of formations is that they can be mapped, not only on the surface, but from photographs taken several miles above the ground. Fig. 2-10 shows an aerial photograph of an area and Fig. 2-11, the corresponding geologic map. Because different formations are composed of different rock types, they commonly have different colors and topographic expressions, and they weather to different soils that support different kinds of vegetation. These things show up well on aerial photos. In mapping, the focus is placed on the contacts between formations because these become lines on the map. Each formation, then, is expressed by a particular pattern or color between two contacts. In the field the area indicated is underlain by the rock unit depicted for it. Geologic maps are objective because the units expressed on them are distinct and recognizable, and can be distinguished from adjacent units in the field.

Geologic maps do not show time lines. These can only be inferred by detailed work on the outcrops themselves, and they are not expected to coincide with the mapped lithologic contacts. The area shown in Fig. 2-11 is small, and if time lines could be drawn on it, they would diverge only slightly from the formational contacts shown. Precisely how they diverged, however, would reveal the directions the environments migrated, necessary information in interpreting the area's geologic history.

FIG. 2-10 Aerial photograph of an area in Larimer County, Colorado, showing sedimentary rocks (center and right) and metamorphic rocks (light curved segment on left). Dip of the sedimentary rock units is to the east (right). (Courtesy Agriculture Stabilization and Conservation Service, USDA.)

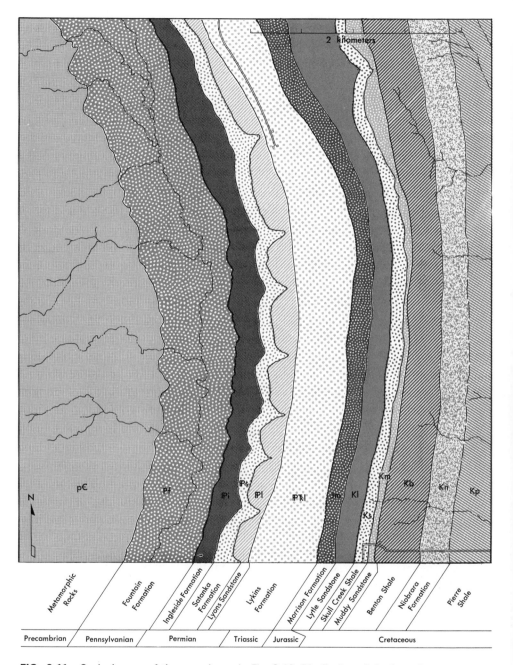

FIG. 2-11 Geologic map of the area shown in Fig. 2-10. Distribution of the formations was determined solely from the aerial photograph in Fig. 2-10. The nature of the formations was verified by field work on the ground. Ages in legend were assigned by correlating the formations with rocks of known ages elsewhere.

Geologic Cross Sections

Geologic maps and aerial photographs give one a vertical or "plan" view of an area. A side view of an imaginary vertical slice through the crustal rocks is also helpful in conveying the configuration of rock bodies. Such a view is called a *geologic cross section*. In addition to providing a third dimension, cross sections generally show greater stratigraphic detail than maps because the vertical dimension can be exaggerated in order to clarify the relationships among rock units.

Relative Dating of Geologic Events

Geologic maps and cross sections help to unravel geologic history. This process begins with the relative dating of rock bodies and of geologic events.

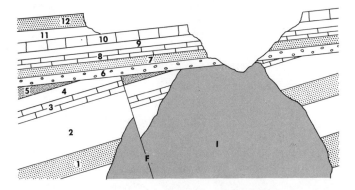

FIG. 2-12 Geologic cross section of an area illustrating two sedimentary episodes separated by igneous intrusion, faulting, and erosion.

Figure 2-12 shows an example. Here an early sequence of strata (1 through 5), an intrusive igneous body (I), a fault (F), and a later sequence of strata (6 through 12) record successive historic events. The evidence for this order is provided by two important principles: *superposition* and *cross-cutting relationships*. The principle of superposition stipulates that, in a succession of strata, the overlying beds are younger than underlying beds. The principle of cross-cutting relationships stipulates that rocks cut by a fault or an intrusive are older than the fault or intrusive. In addition we may infer that the sharply truncated surface overlain by the 6 through 12 sequence was produced by uplift and erosion similar to that which has shaped the present surface. The cross section thus reveals the following history for the area:

1. Deposition of beds 1 through 5.
2. Igneous intrusion.
3. Faulting.
4. Uplift and erosion.

5. Subsidence and deposition of beds 6 through 12.
6. Uplift and erosion.

The Regional Picture

Once the events of an area have been organized into a plausible history, the next step is to fit them into the geologic history of the entire region. To accomplish this one must correlate the rocks of the local area with those of other areas, arrange their records into a composite picture of laterally-equivalent rock types, and establish their ages in terms of the geologic time scale. The fundamental units of the geologic time scale are the geologic periods and their subdivisions. These constitute our standard of time measurement (see the geologic time scale on the last page of this book).

If a given area is near others in which rocks are well dated, it may be possible to correlate them on the basis of key strata and other features they may have in common. But if the area is isolated, there are ordinarily just two methods by which to establish the time when the rocks were deposited. The first, and by far the most widely used, is by fossils. The second is by determining ages of radioactive minerals in years. In either case one must adopt a broader scheme, which transcends local rock units and which seeks to recognize wide-spread units defined on the basis of contemporaneity. The regional picture is thus assembled on the basis of time stratigraphy.

If formations extended worldwide and if their basal and upper boundaries were parallel to time boundaries, then rock stratigraphy and time stratigraphy would be indistinguishable. In tracing rock units, one would be simultaneously tracing time units. In nature two factors combine to complicate this simplistic picture, and these factors create the need for one kind of unit for rocks and an entirely separate kind of unit for time. The two factors are (1) facies and (2) gaps in the stratigraphic record. This chapter will conclude with an examination of these two topics.

FACIES

Just as rock types change vertically due to changing environmental conditions through time, they also change laterally due to differing environmental conditions in space. For example, at a given locality a conglomerate unit 100 meters thick contains about 20 percent interbedded sandstone. These rocks are carefully traced laterally. Within 2 kilometers sandstone beds have become as abundant as conglomerate beds, and within 10 kilometers the unit has become entirely sandstone except for a few siltstone and shaly sandstone interbeds. With further tracing one finds that shale beds become increasingly common, and within a few more kilometers shale dominates the stratigraphic interval that, in the original area, was entirely conglomerate. These different lithologic aspects of the same stratigraphic interval are called *sedimentary facies*, and could be termed the conglomerate facies, the sandstone facies, and the shale facies of the stratigraphic interval in question. Other facies changes may be more subtle. These include changes in mineralogical or chemical composition, texture, stratification, sedimentary structures, or fossil content of the rocks. Lateral changes in physical characteristics may be termed *lithofacies* to distinguish them from changes in the fossil content, or *biofacies*.

To most geologists the term "facies" not only implies that the strata having the different characteristics are laterally contiguous but also that they are the same age. Others, however, sometimes use facies terminology not only for lateral changes but also for vertical changes in lithology or fossil content. Such usage broadens the term "facies" to refer not only to lateral aspects of particular stratigraphic units but also to *lithotopes*—the rock record of particular sedimentary environments without regard to their lateral, or for that matter, their vertical, relationships. In reading of "facies," then, one must note the sense in which the term is used. Usually, an author's meaning is apparent from context.

Laterally different facies result from different coexisting environments. Some of the environments listed in Table 2-2 exist adjacent to one another in nature, but others do not. Hence, facies changes do not occur randomly between any two lithotopes. Instead, they reflect lateral environmental relationships that actually prevail today and that evidently prevailed in the geologic past, too. For example, marine environments rarely border continental environments directly; a belt of transition environments, however narrow, nearly everywhere separates the two. Similarly, deep-sea environments cannot directly border transitional environments like beaches or lagoons; there must be an area of shallow water between them. Thus, not only do continental and marine facies or transitional and bathyal facies never directly intergrade laterally, but also, if deposition is continuous, their deposits cannot directly succeed each other vertically.

If the boundary between depositional environments remained relatively stationary, the resulting lithotopes would maintain their geographic positions and, except for small-scale intertonguing, the boundaries between facies would be essentially vertical. Rarely does this static situation long prevail, because it requires a precise balance between supply of sediment and rate of subsidence. Far more commonly, rates of subsidence and supply differ, and environments migrate, their deposits covering earlier deposits of laterally-occurring environments. Figure 2-13 illustrates relationships in a typical array of environments bordering a seacoast. When subsidence exceeds supply of sediment (Fig. 2-13(A)), the shoreline moves landward in *marine transgression*. Deposits formed in continental environments are overlain by those of transitional environments, which are, in turn, overlain by marine deposits. When the supply of sediment exceeds subsidence, the shoreline builds seaward, causing *marine regression* (Fig. 2-13(B)). Regression forms a vertical sequence that is the reverse of transgression.

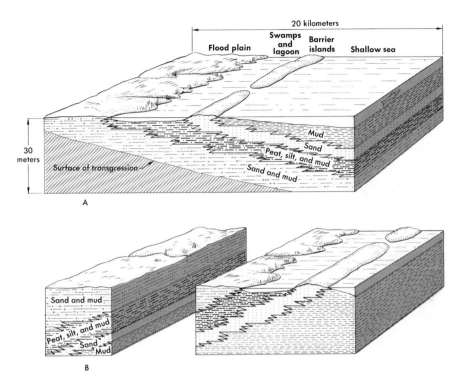

FIG. 2-13 Block diagrams of facies in nearshore environments showing the opposite succession produced by (A) transgression, (B) regression. The vertical scale is exaggerated about 500 times.

Transgressive and regressive sequences generally do not contain a complete sedimentary record of all environments that existed at the time. Transgressive sequences are incomplete if the supply of sediment was not adequate to produce fairly continuous deposition as the sea advanced. Regressive sequences are incomplete where regression was caused by uplift of the land (or lowering of sea level); in fact, erosion typically removes some of the previously deposited sediment. However, in the San Juan Basin of New Mexico, depositional transgression and depositional regression of the Late Cretaceous sea produced nearly complete cycles. Exposures are excellent, and individual sandstone and shale tongues at facies boundaries are locally thick enough to be mapped. Figure 2-14(A) shows that the thick Cliff House Sandstone, which in interpreted as a transgressive barrier-island deposit (like that shown in Fig. 2-13), intertongues southwestward with the underlying lagoonal deposits of the Menefee Formation and northeastward with the overlying marine Lewis Shale. Thus, even in a small area, the migration of marginal and marine facies through time is clearly apparent. One could never mistake the intertonguing contact of the Cliff House Sandstone and Lewis Shale in Fig. 2-14(A) as a time surface.

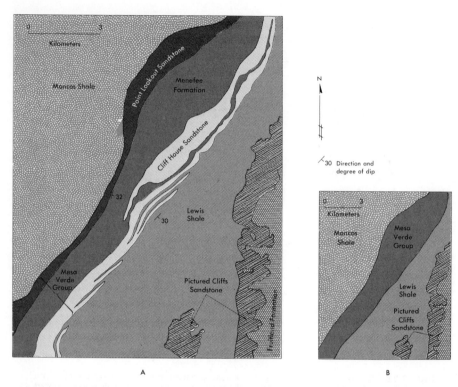

FIG. 2-14 Two geologic maps of the same area of Cretaceous rocks in western San Juan Basin of New Mexico. Map A shows intertonguing of the Cliff House Sandstone with underlying Menefee Formation and overlying Lewis Shale. (After a map by O'Sullivan and Beaumont, 1957.) Map B is drawn at too small a scale to show intertonguing, and top of the Cliff House (top of the mapped Mesa Verde Group here) appears as a smooth line. (After a map by J. B. Reeside, Jr., 1924.)

Although the intertonguing of rock units is extremely common, it can rarely be delineated on a map as in Fig. 2-14(A) because individual tongues of adjacent facies are ordinarily much thinner than those of the Cliff House. Even the thick Cliff House tongues could not be easily mapped except where exposures were excellent, and even then, they could not be mapped at a substantially smaller scale. Figure 2-14(B) is taken from an older map of exactly the same area as Fig. 2-14(A), but it was made at just half the scale. On this map it simply was not possible to show the intertonguing detail. The point is that this type of representation is the rule rather than the exception. Inasmuch as intertonguing of rock units cannot be shown on geologic maps, it is rarely possible from maps alone to determine the direction in which environments migrated. Intertonguing relationships must almost always be depicted with the aid of cross sections on which vertical exaggeration is possible.

GAPS IN THE RECORD

Beside facies changes, gaps in the stratigraphic record constitute the other major factor that complicates the working out of time stratigraphy. In one sense the gaps in the stratigraphic succession are more difficult to recognize and evaluate than facies changes because evidence for missing strata is necessarily negative. Large gaps, called *unconformities*, represent substantial missing portions of the geologic record, and these are important in working out geologic history. Small stratigraphic gaps, which are sometimes called *diastems*, are generally less important, but they are vastly more numerous. Hence an understanding of them provides significant insights into the nature of the stratigraphic record.

Diastems

In Chapter 1 we saw how some geologists around the turn of the twentieth century attempted to quantify geologic time based on estimates of rates of deposition of sedimentary strata. A short time later it became possible to reverse the process and to estimate rates of deposition of ancient sedimentary rocks from the newly available radiometric ages. Joseph Barrell in 1917 pointed out that these estimates posed a problem because they indicated that strata deposited since the beginning of the Paleozoic Era, even where they were thickest, had accumulated much more slowly than anyone had previously inferred—thousands, rather than hundreds, of years per foot. Yet there is good evidence that many individual strata accumulate rapidly. Barrell concluded that innumerable diastems, most of which probably represent years or hundreds of years, must permeate stratigraphic sequences and must actually account for the bulk of elapsed time.

Barrell utilized the concept of baselevel to explain diastems. Baselevel had been conceived earlier as that imaginary surface *below* which erosion—chiefly stream erosion—cannot occur. Barrell extended baselevel into sedimentary environments where it became the surface *above* which sediments cannot accumulate permanently. The same baselevel is thus a surface of aggradation and one of erosion. A baselevel, in spite of its name, is not absolutely horizontal; rather it tilts gently basinward and may even undulate. A familiar example of baselevel is the depth in a shallow sea of the effective wave base. Above this point, wave energy prevents the permanent accumulation of sediment by moving deposits periodically until they ultimately come to rest in deeper water. In areas of erosion baselevel is the surface at which streams must cease cutting downward and below which they cannot erode. The position of this surface results from a complex interplay of several factors, one of the most significant being climate.

Areas of erosion, by definition, lie above baselevel, and areas of deposition below. Hence, between the two, where neither erosion nor deposition occurs, baselevel must intersect the Earth's solid surface. If all factors remained constant, the baselevel surface at which forces of aggradation and erosion are balanced would be stationary. In nature, however, environmental factors never remain altogether constant for long; thus baselevel is subject to short-, intermediate-, and long-term fluctuations.

In shallow seas baselevel is largely governed by sea level. If sea level rises, waves cannot erode so deeply and streams cannot cut so deeply on adjacent land. Thus, baselevel likewise rises and moves landward. If sea level drops, waves and streams erode more deeply; baselevel drops correspondingly and moves seaward. If sea level remains constant, the only way for sediment to accumulate is for subsidence to occur. The total amount of subsidence is the chief factor governing how thickly sediments can accumulate in an area.

Barrell viewed all factors that influence baselevel—basin subsidence, climate, sediment supply, wave activity, and so on—as being intermittent or cyclic rather than constant. Some of the cycles are large; others are small. At times their effects offset one another; at other times they are additive. Baselevel therefore constantly fluctuates; but ultimately, areas that have net subsidence preserve a net accumulation of sediments. This concept explains both rapidly deposited individual beds and a vastly slower average rate of sediment accumulation. Deposition typically occurs in pulses, and the bedding planes commonly represent more time than the beds.

Figure 2-15 graphically portrays this concept. A hypothetical area subsides by 200 meters and hence receives 200 meters of sediment, shown in the column on the left side of the figure. Superposed on the major cycle of subsidence, however, are smaller cycles representing two lower orders of magnitude. Perhaps these are sea level changes and climatic changes respectively. There must be a net accumulation since the overall setting is one of subsidence, but the smaller cycles sometimes add to the major trend of accumulation and at other times subtract from it, causing temporary episodes of erosion. When baselevel, as shown by the graph, moves up, accumulation occurs; when it moves down, erosion occurs. Only the small fragments of elapsed time shown by the white bars at the top of the figure are actually represented in the sedimentary column on the left. Although this diagram is not an actual case, the principle it expresses is believed to be prevalent in most depositional environments, except that cycles in nature are probably more irregular.

Relative subsidence is necessary for the permanent accumulation of sediment, but subsidence alone is no guarantee that accumulation will occur. Without an adequate supply of incoming sediment, subsidence may simply create deep oceans or lakes. Even in the deep oceans, normally thought of as being far below baselevel, significant erosion may occur, chiefly by solution of chemical sediments, but in some cases by the scouring action of deep currents.

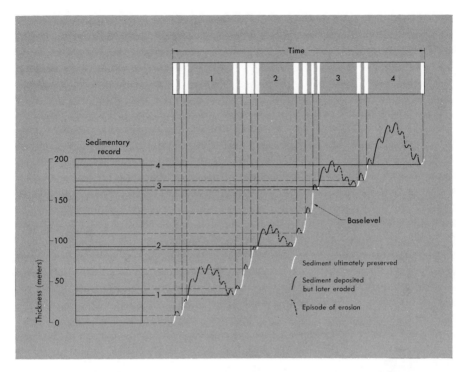

FIG. 2-15 Intermittent deposition in a sedimentary basin. Major trend is one of aggradation, but second- and third-order cycles create a discontinuous record. Where the baselevel rises, sediment accumulates; where it falls, erosion occurs. White portions of the curve are ultimately preserved in the sedimentary record. Total amount of time represented in the stratigraphic record at left is shown in white bars at top. Major diastems are numbered. (After J. Barrell, 1917.)

Unconformities

An unconformity is a large gap in the geologic record formed when deposition ceased for a considerable time. It almost always results from uplift, which causes the erosion of some of the previously formed record. The magnitude of the unconformity is measured by the total span of time for which there is no sedimentary record. Wherever sedimentary sequences have been studied, they are interrupted by such gaps. The lost intervals range in magnitude from the smallest time-stratigraphic unit that we can recognize to entire eras. The time represented by a given unconformity varies laterally, commonly decreasing toward the sedimentary basin that received the material that was eroded during the production of the unconformity. Here the unconformity may disappear entirely into the midst of a sedimentary sequence.

In the field unconformity surfaces are mapped in the same way as any other contacts. Four kinds of unconformities may be distinguished, based on the structural relationships of the units above and below the surface (see Fig.

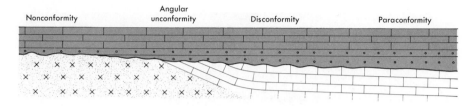

FIG. 2-16   Kinds of unconformity.

2-16). *Angular unconformity* (Fig. 2-17) represents the folding of an older
sedimentary record, its planing down by erosion, and the deposition of a younger
sequence that truncates the old. *Nonconformity* (Fig. 2-18) refers to a surface
at which stratified rocks rest on intrusive igneous rocks or metamorphic rocks
that contain no stratification. *Disconformity* (Fig. 2-18) refers to an unconformity in which the beds above and below the surface are parallel. Disconformities
show some erosional relief, solution features, phosphatic nodules, or other
physical evidence of a break in the sedimentary record. On the other hand,
some unconformities in parallel strata that represent a large time gap over

FIG. 2-17   Angular unconformity between steeply dipping Silurian beds below and
gently dipping Old Red Sandstone above at Siccar Point, Scotland, where James
Hutton first recognized the historical significance of unconformity.

FIG. 2-18 Grand Canyon section looking northwest from Hopi Point, showing the nonconformity (N) between Precambrian igneous and metamorphic rocks and overlying Cambrian Tapeats Sandstone, and the disconformity (D) between Cambrian Muav Limestone and overlying Mississippian Redwall Limestone. The vertical lines in the Redwall are caused by joints. Bedding is horizontal. (Courtesy W. C. Bradley.)

vast areas look just like ordinary bedding planes. Such unconformities appear conformable in the field, and they have been termed *paraconformities*. Paraconformities are recognized chiefly on fossil evidence. Where there is no fossil evidence for a significant time break, some could—and undoubtedly do—go undetected.

In addition to varying in time value, the same unconformity surface may change from one kind of structural relationship to another when traced laterally.

What appears as a featureless disconformity in one area may have discernible relief in others. An apparent parallel relationship in a small area may be shown to be regionally angular, with rocks that range greatly in age being truncated beneath the surface. The unconformity may even be traceable into areas where angularity is clearly apparent. Degree of angularity or amount of relief on the unconformable surface bears no relationship to an unconformity's time value.

In nature, all gradations occur between the smallest breaks at bedding planes, which may represent no more than the time between two big storms, and the largest unconformities. Nevertheless, it is convenient to recognize the distinction between diastems that are normal components of *conformable* sedimentary rock units and large *unconformable* breaks during which there were fundamental changes. Unconformities usually represent so much time that when deposition reappears, an overall change in the environment has occurred. Diastems do not, and this is the basic difference between them. Viewed in terms of baselevel movements, diastems result from minor fluctuations as baselevel moves generally upward, whereas unconformities result from major long-term regressive-transgressive cycles in which baselevel first moves downward relative to the Earth's surface to produce nondeposition and erosion, and later upward again to re-establish depositional conditions.

An unconformity produced by a major regressive-transgressive cycle is shown in Fig. 2-19. In the geologic cross section at the top, the sedimentary sequence "A" rests unconformably on igneous rocks, and is overlain unconformably by sedimentary sequence "B." The lines A_1 through A_7 and B_1 through B_3 are time lines (indicated by T_0 through T_{10} at the right margin). The bottom diagram represents the same geologic record but its vertical dimension is *time* rather than thickness, so the time lines are horizontal. Deposition of the "A" sequence continued across the entire area nearly until time T_5 when horizon A_5 was produced. Then erosion began. The change from deposition to erosion at the top of sequence "A" was caused by a lowering of baselevel relative to the Earth's surface. As baselevel moves downward, its point of intersection with the Earth's surface moves slowly basinward to the right along line MP in the bottom diagram of Fig. 2-19. At any point on line MP where baselevel intersects the Earth's surface, such as point C, erosion is in progress to the left and deposition is in progress to the right. At time T_6 when the baselevel-surface intersect is at point C, erosion has removed the record to the level of dashed line E_6. The missing record above line MP is due to nondeposition. The process continues to point P at Time T_7, at which time erosion has progressed to the level of E_7.

Here the cycle reverses; baselevel moves upward and landward. Deposition to the right of point P continues uninterruptedly and deposition to the left is progressively re-established. The baselevel surface intersect has arrived at point D by time T_8, and erosion has progressed to surface E_8. Finally, at time T_9 erosion has ceased at the surface LP. The entire area is again one of deposition

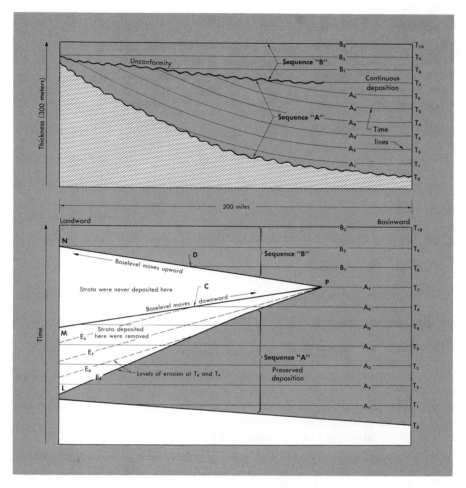

FIG. 2-19 (Top) Geologic cross section showing a sedimentary sequence "A" resting non-conformably on igneous rocks and overlain unconformably by another sequence "B" (vertical exaggeration about 300 times). (Bottom) Analysis of unconformity between "A" and "B" with vertical scale as time rather than thickness. (Modified from H. Wheeler, 1964.)

and the depositional surface NP is the same as erosional surface LP. This is the surface of unconformity. The total time represented by the unconformity clearly increases to the left from point P, and the parts of the record missing due to both nondeposition and erosion show clearly.

three

the time-stratigraphic record

The first clear statement that layered rocks show sequential changes—that they have histories—was made by Nicolaus Steno in 1669. From his work in the mountains of western Italy Steno realized that the principle of *superposition* in the stratified rocks was the essential key. Steno also realized the importance of another principle—*original horizontality*—namely, that strata are always initially deposited nearly horizontal although they may be found dipping steeply. Nearly one hundred years later in 1760, Giovanni Arduino classified the rocks of the same region into three main categories:

1. Primary—crystalline rocks with metallic ores.
2. Secondary—hard stratified rocks without ores but with fossils.
3. Tertiary—weakly consolidated stratified rocks usually containing numerous shells of marine origin. Volcanic rocks were included in this division.

In a fourth, less important category he recognized alluvium washed from the mountains and lying on the plains.

Almost simultaneously, Johann Gottlob Lehmann in 1756 and Georg Christian Füchsel in 1761 were separately applying the principle of superposition to two areas in Germany. Lehmann reconized three categories of rocks—crystalline, stratified, and alluvial—a classification similar to Arduino's. Füchsel worked only with the middle category—the stratified rocks—which he divided into nine rock units, but he made it clearer than either of his contemporaries that rock units represented discrete episodes of geologic time. A short time later, in 1777, Peter Simon Pallas recognized a threefold sequence of primary, secondary, and tertiary rocks in the Ural Mountains of central Russia.

The principle of superposition and recognition of a threefold division of rocks based upon it thus came to be widely accepted in Europe in the last part of the eighteenth century (see Fig. 3-1). This provided a basic framework for stratigraphic thinking and directed attention to the time relations of strata.

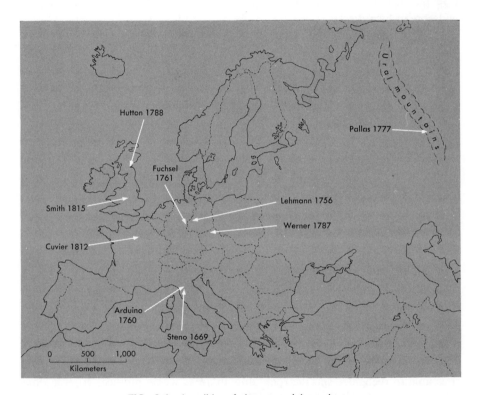

FIG. 3-1 Localities of pioneer work in geology.

The threefold stratigraphic divisions in northern Germany, central Russia, and western Italy surely represented time spans of Earth history, and workers began to wonder if they represented the same time spans. Suppose they did, then what about the numerous thinner individual rock units within each main time division? Did they likewise have synchronous lithologic counterparts in the diverse areas?

The answer is that individual rock units almost never extend so widely. This could be established by determining that rock units which we would call "formations" are not continuous between Germany, Italy, and central Russia. Units of similar lithologies may occur, but they were never connected and their order of sequence is certain to be different. In the last part of the eighteenth century these facts were unknown. The detailed work necessary for an answer simply had not been done. In this atmosphere of uncertainty Abraham Gottlob Werner, who had never seen the rocks very far beyond the boundaries of his native Saxony, made his famous pronouncement that, indeed, individual rock units were universally distributed. As we have seen, Werner's Neptunist scheme held sway for several decades, and an important factor in its eventual collapse was the realization that stratification sequences differ significantly among widespread localities.

When the Wernerian scheme died, everyone realized that it could no longer be considered possible to know the stratigraphic sequence of the world, or even of a single region, by examining the strata at just one place. Stratified rocks therefore seemed to pose more problems than ever. The need for detailed mapping of rock distributions was becoming increasingly apparent, but detailed geologic mapping in a humid region like Europe is difficult. Almost everywhere the rocks are covered superficially by soil, vegetation, or alluvium. Even if one could readily map rock units in local areas correctly, one could rarely be certain of relationships between widely separated areas of outcrops. The need at this juncture was for a new unifying principle—a new tool—by which units could be categorized and recognized widely.

William Smith's Discovery

William Smith (1769–1839), working in England, found in fossils just this tool. Smith was an engineer and surveyor, and his investigations for roads, quarries, mines, and canals acquainted him intimately with much of England's countryside. During his travels, Smith recognized and traced out numerous sedimentary rock units, and he soon noticed that each successive unit contained its own diagnostic assemblage of fossils by which it could be distinguished from other units of different ages. For different units of similar lithology this was commonly the only way certain identification could be established. Smith was thus able to recognize specific rock units in areas where physical criteria alone were inconclusive. Utilizing his new principle, he produced in 1815 the first geologic map of England, Wales, and part of Scotland. This large and attractive map, more accurate than any attempted previously, represented a landmark in the representation and, hence, the understanding of rocks. Smith's discovery that strata may be identified by the fossils they contain shortly thereafter became known as the *law of faunal succession*. This important principle raised questions concerning ancient life that were not to be answered easily, but even without the answers to these questions, correlation between distant localities now became feasible. The way was prepared to erect a stratigraphic classification based on time relations of strata rather than on rock types. In short, this was the key discovery that stratigraphic geology needed to progress further.

DISCOVERY OF ORGANIC EXTINCTION

In the Middle Ages fossils were considered to be inorganic "figured stones" or "sports of nature." Some attributed them to an occult power at work in the Earth; others held that they were "irradiations" from the stars and planets. Leonardo da Vinci, in 1508, was among the first who recognized that fossils were the remains of seashells and other animals just like those living today. This view eventually gained sufficiently substantial support and, fortunately, by Smith's time, fossils were clearly understood to be the remains of once-living organisms.

The considerably more radical idea that some formerly existing species were now extinct—an idea that had been postulated long before Smith's time—had simply not been proven to the satisfaction of most people. The chief reason was that the best-known fossils were marine invertebrates. These are certainly as good as any animals to demonstrate extinction, but at the beginning of the nineteenth century little was known of the present life of the oceans. Mindful of the very real limits of their knowledge, naturalists of the time hesitated to suggest that certain marine animals represented by fossils no longer existed anywhere on the Earth.

Following the lead of Smith, Georges Cuvier (1769–1832), a French zoologist, worked out the stratigraphic sequences of terrestrial vertebrates as well as the marine invertebrates in the Tertiary strata of the Paris Basin. In 1812, he showed conclusively that many fossil vertebrates have no known counterparts living today, and everyone considered it highly unlikely that such big land animals could be undiscovered. Extinction was at last a reality; it was then a short step to recognize the principle for invertebrate animals and plants as well. Cuvier carefully worked out a succession of different faunas in the Paris Basin strata, and he noted that the younger deposits contained creatures more like those of the present day than did the older deposits. Cuvier felt that the rocks thus revealed advancing complexity of life.

The strata of the Paris Basin, like those of many areas, represent a very incomplete record, so Cuvier was unable to find gradational forms connecting the different faunal levels. Instead his attention focused on the stratigraphic breaks between them, breaks that were evidenced by abrupt faunal changes, abrupt lithologic contacts, and sometimes by conglomerate beds. Impressed by these breaks in the stratigraphic record, Cuvier postulated that each represented a sudden, horrendous worldwide catastrophe that destroyed the prevailing life of the time whether terrestrial or marine. After each catastrophe a new fauna was created, and it prevailed until its destruction in the next catastrophe. Popular opinion quickly assumed that the last catastrophe of all was the Biblical Deluge and that the earlier epochs of life were somehow equated with the Biblical Days of Creation.

FOUNDING OF THE GEOLOGIC SYSTEMS

Smith showed that sedimentary rock units may be identified by their distinctive fossils. Cuvier went a step further in emphasizing the changing sequence of faunas, each representing a particular age, and he established the reality of the extinction of species. Although geologists have long since dismissed his interpretation that each fauna appeared following a tumultuous catastrophe, Cuvier refined the tool of faunal succession sufficiently so that rocks of the same age could be recognized in far-flung places. Following the work of Smith and of Cuvier, it became possible to define major units of sedimentary rock and, using their distinctive fossils, to distinguish their time counterparts on opposite sides of an ocean if necessary, even where rock types differ. These major stratigraphic units, each of which represents an important segment of the geologic time scale, became known as the *geologic systems.*

The geologic systems in use today were not proposed by one man, nor were they the suggestion of a learned committee. They grew up without plan through the efforts of numerous geologists working independently. In the process, different schemes for subdividing the sedimentary rock record were suggested. Some of these were quickly taken up and utilized by later workers; others were never used again and have been forgotten. Today's geologic time scale has evolved through trials by numerous geologists in the actual work on the rocks over the decades. It has been quite stable since the turn of the twentieth century, but it is still subject to future refinements. Most of the presently recognized geologic systems were defined in Europe. The rock outcrops that actually constitute the basis for definition of a system are called its "type section," and the area in which they occur is called the "type area." We will briefly review how the systems that are presently used became established.

Cambrian, Ordovician, and Silurian

Study of the complex geology of western Wales by the British geologists Adam Sedgwick and Roderick Impey Murchison resulted in 1835 in the naming of the Cambrian and Silurian Systems for ancient Welsh tribes. Each worker attempted to recognize breaks in the stratigraphic record as boundaries for his subdivision. Murchison began with the top of the sequence in the southeast, and Sedgwick began at the base in the northwest, and they worked toward one another. Murchison carefully documented the abundant fossils of his Silurian strata, but Sedgwick's strata were poorly fossiliferous, and his breakdown of the Cambrian System was essentially lithologic. When it became clear that their systems overlapped, a quarrel ensued. The controversy was not resolved until 1879 when Charles Lapworth proposed the name "Ordovician System," taken from that of another Welsh tribe, to include the disputed interval between the Cambrian and Silurian, and to express a threefold paleontologic division in Early Paleozoic strata of Europe that had by then become apparent. Boundaries of Lapworth's Ordovician System were solely paleontologic.

Devonian

In 1840 Murchison and Sedgwick jointly named the Devonian System for the rocks of Devonshire in southern England, having done the actual work prior to their misunderstanding over the contact between the Cambrian and Silurian. Devonshire is a poor type area because the rocks are intensely deformed and the base of the system is not exposed. Nevertheless, the rocks are fossiliferous, and it was their distinctive faunas—intermediate between those of the Silurian below and the Carboniferous above—that led to their identification as the Devonian. Murchison and Sedgwick showed that fossils could be used to recognize the Devonian System in the Rhineland where it is much better exposed and much more fossiliferous.

Carboniferous

Two British geologists, William Conybeare and William Phillips, proposed this name in 1822 for the strata in north central England that contained coal beds. The Carboniferous System was one of the first to be proposed following recognition of the value of fossils for correlation. The term "Carboniferous" ("coal-bearing") is descriptive, but Conybeare and Phillips expected that the Carboniferous System would be widely recognizable by its distinctive fossils rather than its lithology.

Mississippian, Pennsylvanian

Alexander Winchell introduced the term "Mississippian" into American stratigraphic terminology in 1870 for the well-exposed Lower Carboniferous strata of the Mississippi Valley. In 1891 Henry Shaler Williams coined the name "Pennsylvanian" from the state of Pennsylvania for Upper Carboniferous strata as a counterpart to Winchell's Mississippian strata. T. C. Chamberlain and R. D. Salisbury elevated both terms to system status in their influential geology textbook of 1906, and they justified this division largely on the basis of the supposed widespread unconformity that separated the two. The U.S. Geological Survey has recognized these systems officially only since the mid-1950's. Neither has found use outside North America.

Permian

Murchison in 1841 named this system from the province of Perm in Russia, where it consists of a great thickness of limestones overlying the Carboniferous strata. Murchison recognized that the fossils differ from those found in older and younger strata, and he ascertained that Permian strata could be identified in widespread areas by their distinctive fossils before he formally named this system.

Triassic

Friedrich von Alberti, a longtime official in the German salt-mining industry, introduced this name in 1834 for a sequence of rocks with a striking threefold division. The term is thus essentially descriptive. In the type area of southern Germany the strata are widely traceable but poorly fossiliferous. In the Alps to the south, a complete sequence of marine faunas today provides the standard of reference for worldwide correlation.

Jurassic

Alexander von Humboldt, a pioneer German geologist, first coined this term in 1795 for the strata of the Jura Mountains in northern Switzerland, but at this early date he considered it only as another "formation" in Werner's Neptunist scheme. Leopold von Buch in 1839 redefined the Jurassic as a system in its own right.

Cretaceous

A Belgian geologist, Omalius d'Halloy, in 1882 proposed this term for strata encircling the Paris Basin. The term derives from the Latin word for chalk and is entirely descriptive. As used today the Cretaceous System includes more than just chalk beds, even in its type area where the lower portion contains chiefly sandstone and shale. In other parts of the world the system may be thousands of meters thick and include no chalk beds at all.

Tertiary

This is the only one of Arduino's (1760) terms still in general use. Originally defined in Italy, its constituent series—Paleocene, Eocene, Oligocene, Miocene, and Pliocene—have their type sections in France. The Eocene, Miocene, and Pliocene Series were defined by Charles Lyell (1833) on the basis, not of lithology, but of the relative proportions of the living and extinct fossils each contained: Eocene contained 3 percent living species, Miocene 17 percent, and Pliocene 50 to 67 percent. August von Beyrich (1854) later added the Oligocene, and Wilhelm Schimper (1874) added the Paleocene Series.

Quaternary

In 1829 a French geologist, Jules Desnoyers, proposed this term in France for very young strata. Today it includes the Pleistocene Series (proposed by Lyell in 1839), which constitutes deposits formed during the glacial ages, and the

Recent Series (proposed by Lyell in 1833), which is a rather poorly defined term for any postglacial deposits.

Summary of the Founding of the Time Scale

Type regions for those systems named in Europe are indicated in Fig. 3-2; this map includes all commonly used systems except Mississippian and Pennsylvanian, which are widely used in North America. Other system names are in limited use. For example, some geologists in northern Europe use "Gotlandian" in place of Silurian in its modern sense. In place of the Tertiary many stratigraphers now employ the terms "Paleogene" and "Neogene," using the top of the Oligocene as the dividing boundary.

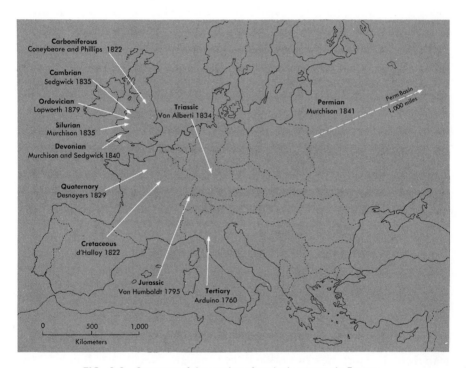

FIG. 3-2 Summary of the naming of geologic systems in Europe.

This brief sketch of the founding of the time scale shows above all that the geologic systems were delineated without benefit of any grand design. Some were established on distinctive lithologies, some on distinctive faunal content, and some on major breaks in the stratigraphic record. Regardless of how the systems were conceived, however, each has been subsequently discriminated and utilized on every continent on the basis of distinctive fossils. The systems in use today have by and large proven to be remarkably convenient units, and their worldwide utility is the reason that the geologic time scale, as outlined here, has survived.

TIME-STRATIGRAPHIC UNITS AND
GEOLOGIC TIME UNITS

During the nineteenth century, while geologists were naming the geologic systems, they were simultaneously developing a concept of geologic time. The time framework is based upon the systems, each of which—in its type area at least—is represented by a considerable thickness of strata. Each system lies between older strata below and younger strata above, and the systems are contiguous; that is, they do not overlap in time. Where one stops in the stratigraphic succession, another begins. Thus any given system was deposited during a discrete portion of geologic time. Any rock, anywhere, which was formed within that time range belongs to that particular system. But the actual geologic time represented by each of the systems of course occurred everywhere in the world, even if rocks representative of the system are missing as a result of nondeposition or later erosion.

Rocks formed during a specified interval of time are called *time-stratigraphic units*, and the geologic systems are the basic time-stratigraphic units of historical geology. A time-stratigraphic unit is independent of rock types or thicknesses. It cannot be recognized by rock type except in one place—the type locality—where it exists by definition. Everywhere else it must be recognized by time correlation. The type section of the Ordovician System in Wales, for example, has been defined as a particular body of strata. From the kinds of sedimentary rocks and from the fossils they contain, we can read a local history. Strata immediately below and above the Ordovician record the Cambrian events that preceded and the Silurian events that followed. The rocks in question belong to the Ordovician System without any doubt because they have been defined that way.

Suppose we now inspect a stratigraphic section in the central United States. We can discriminate the successive rock types, recognize local rock units, study their faunas, interpret the sequence of environments they represent, and map them. In short, we can work out the complete local history. If now we can determine, through time correlation, that a portion of this section was deposited contemporaneously with the type section of the Ordovician in Wales, we can call these Ordovician rocks. Thus, in using the name "Ordovician," we are saying, "These rocks were deposited within the same time interval as that body of strata in Wales which constitutes the type."

To recognize time-stratigraphic units, therefore, one must establish correlations either directly with the type section or with intermediate sections, which in turn have been correlated with the type. We have several ways of correlating rocks, all of which are subject to some measurement error. By far the most reliable of these methods, especially over long distances, is the use of *fossil zones*—strata that contain diagnostic fossil species.

The abstract *geologic time units* that correspond to the systems are the *geologic periods:* thus we can say that the rocks of the Devonian System were deposited during the Devonian Period. Time terminology is useful in referring to historical events and circumstances. We might wish to state that fish were abundant during the Devonian Period (because their remains are abundant in rocks of the Devonian System). Or, if the Devonian System is absent in a given area, we might seek evidence of whether the Devonian Period was a time of nondeposition or whether rocks were deposited but later eroded away. Of course, we are aware that Devonian time ever existed only because we have a rock sequence formed during that time somewhere else. In clearly conveying inferences made from the actual rocks, we need the abstract time terms as well as time-stratigraphic terms.

Systems are divided into *series*, commonly used with Lower, Middle, and Upper. The parallel time divisions are *epochs*, commonly used with Early, Middle, and Late. For example, the Devonian System, which was deposited during the Devonian Period, is commonly divided into the Lower Devonian, Middle Devonian, and Upper Devonian Series, whose rocks were deposited during the Early Devonian, Middle Devonian, and Late Devonian Epochs (see Table 3-1). Epochs and series are further subdivided into *ages* and *stages*. Like the systems, many of the series find essentially worldwide application. Stages, because they are smaller units, generally can be recognized only within single continents or regions, but some stages can be traced much more widely.

The geologic periods are lumped into larger geologic time units called *eras*. From names proposed by John Phillips in 1841, we now place the periods in the Paleozoic, Mesozoic, and Cenozoic Eras. These names translate literally into "ancient life," "medieval life," and "modern life," and the divisions reflect the profound changes that occurred in living things at the end of the Permian and Cretaceous Periods.

The old rocks below the Paleozoic have always been known widely and simply as the "Precambrian." One trouble with this term is the lack of a complementary term for all subsequent rocks. We can say "Post-Cambrian," which isolates the Cambrian, or "Post-Precambrian," which sounds a bit ridiculous. Nevertheless, the distinction is fundamental, for it separates old rocks without diagnostic fossils—and hence without the criteria for long-range correlation— from younger rocks whose fossils permit long-range correlation—and hence subdivision into smaller time-stratigraphic units. These two great divisions of time may be called *eons*, and for them G. H. Chadwick in 1930 coined the terms "Phanerozoic" (evident life) and "Cryptozoic" (hidden life). The latter term will probably never replace "Precambrian," but the former term has found increasing use.

Table 3–1 Commonly Used Stratigraphic Units with Examples

Time-Stratigraphic Units	Corresponding Geologic Time Units	Biostratigraphic Units	Rock Units
	Era (Mesozoic Era)	Biostratigraphic zones (*Baculites reesidei* Zone)	Group (Dakota Group)
System (Cretaceous System)	Period (Cretaceous Period)		Formation (South Platte Formation)
Series (Upper Cretaceous Series)	Epoch (Late Cretaceous Epoch)		Member (Plainview Sandstone Member of the South Platte Formation)
Stage (Campanian Stage)	Age (Campanian Age)		Bed (Mostly informal terminology such as "the third coal bed")
Time-stratigraphic zone (*Baculites reesidei* Zone)			

Boundary Problems

There are still problems of refinement of the time scale, and most of these are concerned with precisely where the boundaries of the systems (and their subdivisions) should be placed. Boundaries of many of the systems were chosen originally at unconformities. As geological data accumulated it became clear that these gaps were represented by complete stratigraphic successions in other regions. Problems arose as to whether the "excess strata" should be placed in the overlying or underlying system. Many unrewarding searches have been conducted to find the *real boundary* between, say, the Silurian and Devonian, the fallacy being that some sort of natural boundary exists. In fact, the systems were invented for convenience and the boundaries between them, like the boundary between night and day, are gradational. What is needed is an agreed upon boundary that is clearly defined and readily recognized in most places.

Unfortunately, even boundaries of many smaller time-stratigraphic units that are not bounded by unconformities are subjects of dispute. In retrospect this is inevitable because stages, for example, are traditionally defined at "type localities" where they are well exposed, and these type localities are widely scattered in many different regions. Thus when two stages, theoretically adjacent in time, are defined in distant regions, it is extremely unlikely that the top of the older stage in its type area will correspond exactly with the base of the younger stage in its type area. As a result the time-stratigraphic scale that makes up the workaday world of stratigraphy is full of innumerable overlaps and gaps between what are supposed to be contiguous units. Much time and energy have been wasted on arguments concerning in which stage the overlap or gap, when it is discovered, belongs.

The solution proposed for this problem appears simple: abandon stratotypes as representative of time-stratigraphic divisions. Instead carefully select stratigraphic sections where sedimentation seems to have been as continuous as possible, where there are no marked lithological changes, and where there are unbroken evolutionary lineages in several different groups of fossils. Then—and this is important—define only *one boundary*, preferably the lower, of each time-stratigraphic division in this section, and carefully mark the chosen horizon. This method has the advantage that the base of one division then automatically becomes the top of the division below. Even if there is a stratigraphic break at the level chosen, the definition will stand because the missing strata below will automatically belong to the lower division. This recommendation now has wide support and a beginning has been made on the first international agreements concerning placement of critical boundaries. These will provide a firm basis for future correlations worldwide. Although the newly selected time-stratigraphic boundaries will not necessarily make distant correlation any easier than it is now, at least all workers will be striving to correlate agreed-upon boundaries with maximum precision.

BIOSTRATIGRAPHIC UNITS

Stages generally comprise two or more *zones*, bodies of strata that are distinguished by particular fossils. Zones are considered as separate kinds of stratigraphic units called biostratigraphic units. Such units, which are objective, may be interpreted for their significance for time or environment. Those that are interpreted as having time value are, in actual practice, the smallest time-stratigraphic units that can be recognized.

Some geologists, particularly in Europe, employ the term "zone" only for those units that can be interpreted to have time-stratigraphic significance. This is the more traditional practice and its proponents argue that fossil units of temporal value are the only kind worthy of recognition, and hence that they alone justify the term "zone." This approach requires that the temporal value be assessed before a zone ever can be recognized. In this view, the separate category of biostratigraphic units is superfluous, all zones being by definition time-stratigraphic units. The chief difficulty that this approach encounters is this: if one recognizes only the fossil zones that can be considered time-strati-graphic units, then one must first make the *interpretation* and this necessarily becomes part of the *definition*. Other geologists, who would consider zones first as separate biostratigraphic units, say in essence, "Let's *define* zones solely by their fossil content and when we wish to *interpret* one as having time, environmental, or some other significance, we will refer to it in more explicit terms which convey our interpretation." In this view, the biostratigraphic zone is an objective entity that changing interpretations need not alter. The balance of opinion today favors utilizing the separate category of biostratigraphic units so that the process of defining zones may be clearly separated from the process of interpreting them.

Biostratigraphic zones may be defined by: (1) the total stratigraphic range of one species, (2) the stratigraphic range in which one species occurs in abundance, (3) the rocks that contain a particular assemblage of species with no regard to their ranges, or (4) the stratigraphic interval delineated, in one of a variety of ways, by various combinations of overlapping ranges of selected species. Biostratigraphic units will be examined critically in Chapter 5. We should add here only that even though most fossil zones are restricted to single regions, they provide the very backbone of all time-stratigraphic classification in Phanerozoic rocks.

UNCONFORMITY-BOUNDED UNITS

Earlier this chapter pointed out that a complete record of Phanerozoic history can be assembled only from a composite of local sections in which there are no significant stratigraphic breaks. As the time scale becomes more refined it also becomes a valuable tool for investigating the extent and magnitude of

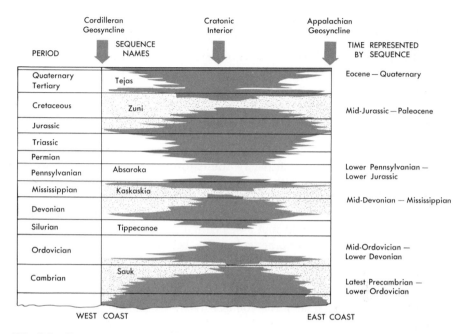

FIG. 3-3 Time-stratigraphic relationships of unconformity-bounded sequences in North America. Dark areas represent large gaps in the stratigraphic record which become smaller toward the continental margins. Light areas represent strata. (Modified from L. L. Sloss, 1963.)

unconformities in various regions. The largest unconformities record continent-wide cycles of uplift and erosion like those diagrammed in Fig. 2-19. Six such episodes have been recognized in the Phanerozoic record of North America. They divide the Phanerozoic into large unconformity-bounded units, which have been called *sequences* in a formal sense. These sequences have been named for convenience in referring to them, but the names are rarely used. Some sequences consist of several geologic systems and others of only part of one system. Fig. 3-3 shows that the sequences represent far more time at the continental margins than they do in the interior where unconformities are largest.

THE PROBLEM OF SUBDIVIDING PRECAMBRIAN TIME

On all the continents the largest exposures of Precambrian rocks occur in vast upland areas called *shields* (see Fig. 6-9). Since Precambrian time, shields have been stable, and they have never been buried deeply by deposits of younger strata. Outside shield areas, Precambrian rocks are exposed in extremely deep gorges, as in the Grand Canyon, and in the cores of mountain ranges. Precambrian rocks differ from Phanerozoic rocks in one important respect—they lack diagnostic fossils. They are mapped the same way and they can be traced locally so long as continuity of outcrop permits. Without diagnostic fossils, however, they simply cannot be correlated between distant areas, and hence time-stratigraphic classification based on systems and series, like that in the Phanerozoic, has not been feasible. The so-called systems and series of Precambrian rocks, such as the "Grand Canyon System" and the "Keewatin Series" of the Lake Superior area, refer not to time-stratigraphic units at all but to large rock units, and they have no application outside their own areas of occurrence.

Numerous schemes for the subdivision of Precambrian rocks have been attempted, and their very multiplicity attests to the fact that none has worked well. During the late nineteenth century the terms "Archean" and "Algonkian" were coined for the Precambrian rocks of the Great Lakes area, and these terms were soon transposed to "Archeozoic" and "Proterozoic." Geologists of both the United States and Canada at first assumed the Archeozoic and Proterozoic to be widely applicable, even to other continents, because prevailing opinion at the time held that (1) the Earth had cooled quickly from a molten condition early in its history, at which time most granite originated, and (2) comparatively little time separated that event from the Cambrian Period; hence no serious error of correlation was possible in the younger, relatively undeformed Precambrian rocks.

The Archeozoic consisted of *basement rocks*—intrusive and extrusive igneous rocks and highly deformed metamorphic complexes as well—thought to date from the early cooling episode. The Proterozoic consisted of overlying Precambrian sedimentary rocks, some slightly metamorphosed. In local areas throughout the vast Canadian Shield region of eastern Canada, counterparts to each subdivision could be identified and numerous rock units belonging to each could be delineated and mapped. The objective local rock units were valid, but the two major subdivisions, insofar as they were supposed to have time significance, were not. Even before the advent of radiometric dating, it became evident that episodes of igneous and metamorphic activity assigned to the Archeozoic were widely separated in time. Radioactive dating methods later showed clearly that many undeformed rocks considered to be Proterozoic were far older than badly deformed rocks that had been placed in the Archeozoic. Radiometric dating thus laid to rest once and for all the idea that rocks can be dated, even in a gross way, by their lithology or by the extent of their deformation and metamorphism.

Radiometric dating also revealed that Precambrian time was far greater than anyone previously imagined. Figure 3-4 shows how the vastness of Precambrian time gradually came into focus as radiometric dating techniques improved. Realization of the magnitude of Precambrian time has been one of the great revelations of historical geology in this century. The Precambrian rocks that we can actually date represent more than 80 percent of the recorded history of our planet. The vast Precambrian terrains, constituting 17 percent

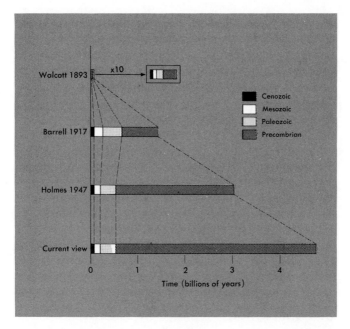

FIG. 3-4 Changing view of the magnitude of Precambrian time in the twentieth century. (After H. L. James, 1960.)

of the Earth's land surface, are monuments to our inability as yet to construct a worldwide stratigraphic scheme without the use of fossils. Any future scheme that may successfully subdivide Precambrian rocks on a worldwide basis will depend on the radiometric dating that has taught us so much already.

four

correlation, stratigraphic mapping, and paleogeography

If, one day in the geologic past, someone had dropped a great number of red marbles close together in a vast area undergoing sedimentation, the sedimentary strata then being deposited would have a built-in layer by which an observer long afterward could recognize the exact level of accumulation on that particular day. Marbles might have fallen on sandy beaches in some areas, on the muddy bottom of a shallow sea in other areas, and on the spongy vegetation of marshes in still others. The marble horizon in the resulting sandstone, shale, and coal strata would represent a plane of time equivalency. Shown in cross section, this horizon would appear as a time line.

Turn to page 53 and picture the surface of a coastal area of sedimentation, like that in Fig. 3-1(A), covered suddenly with the red marbles. With continued sedimentation the marbles would be buried as a single layer. Traced away from the shoreline the marble layer would pass through the various facies forming in the near-shore environments. Traced along any cross section that is parallel to the shoreline, however, the marbles would maintain their relative position within a given lithologic unit. Parallel to the shoreline, in other words, contacts between different lithologies are parallel to time planes inasmuch as the boundaries in this direction alone are forming simultaneously. This direction is called the *depositional strike*.

Although we find nothing so convenient as layers of red marbles in the Earth's sedimentary strata, the goal of our correlation efforts amounts to this: if instantaneously produced layers of red marbles did exist, where would they occur in the sedimentary rocks? Exactly correlative strata were formed at exactly the same time. Correlative strata are bounded by bedding planes, which actually were the Earth's solid surface as it existed at successive times in the geologic past.

Rock Correlation Versus Time Correlation

Some geologists use the term "correlation" to refer also to the simple physical continuity of rock units. As we have seen, this is an altogether different thing from time equivalency. When ambiguity is a danger, always specify "rock correlation" or "time correlation."

In some areas—particularly in subsurface work—bedding continuity, sedimentary structures, and fossils are rarely observed. Geologists rely heavily on physical properties of rocks measured by lowering various instruments into the drill hole. The most widely used of these methods is the electric log, a set of curves that records the rocks' spontaneous potential; that is, the electric field generated by the contact of fluids in the rocks with the drilling fluid, and their

FIG. 4-1 Stratigraphic cross section tracing outcropping Cretaceous rocks eastward into the subsurface of northern Colorado. The topmost line represents a bentonite bed. The other lines represent formation boundaries. Sections 2–5 are 3,000 to 5,000 meters (5,000 to 8,000 feet) below ground surface. (After D. B. MacKenzie, 1963.)

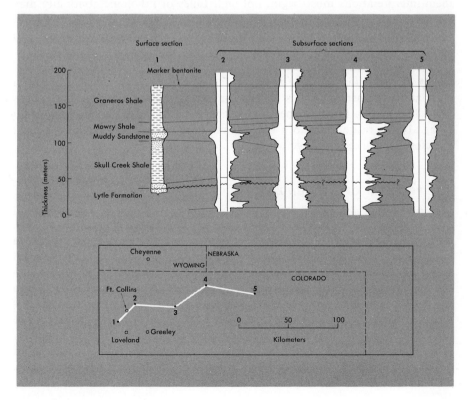

resistivity, measured in an electric field produced artificially within the drill hole. The electric logs in Fig. 4-1 record the electrical properties of a 180-meter sequence of strata in the Denver Basin of northern Colorado, and they make it possible to trace rock units from the outcrop into the subsurface. The electrical properties shown on the logs are largely determined by lithology, which in turn is determined by the environment in which the sediments were deposited. In tracing electrical properties from well to well then, we are likely to be tracing similar ancient sedimentary environments but unlikely to be tracing exact time. Where lines are drawn to connect similar features on electric-log curves, as they are in Fig. 4-1, you should understand clearly that these lines only indicate physical continuity of rock units, nothing more.

It is an old adage in geology that mapping must precede interpretation. In other words, the distribution of the rocks must be correctly observed before the processes that formed them can be properly inferred. In both subsurface and surface statigraphic work the continuity of rock units is depicted best by lines drawn between similar rock contacts on *stratigraphic cross sections*, which are constructed of lithologic or electric-log columns. The lines of lithologic continuity shown must be the starting place for correlation. The important thing to keep in mind is what the lines on stratigraphic cross sections mean. With them interpretations must begin, not end. Lines of lithologic continuity may almost coincide with time lines, but they will not exactly coincide, even over short distances, unless the geographic alignment of the cross section happens to coincide with the depositional strike.

Nevertheless, there are valid physical methods of establishing time equivalency of rocks, just as there are valid faunal methods. The chief problem in correlating sedimentary rocks by any method is that of interpretation across facies boundaries. By the same token, understanding facies depends largely on being able to relate them to a time framework. Facies and correlations must therefore be worked out together.

THE PROBLEM OF RECOGNIZING FACIES

Contacts between adjacent rock units are commonly gradational and the horizon to be mapped must always be chosen somewhat arbitrarily. Between a sandstone below and a shale above, for example, the contact might be selected at the highest sandstone bed a foot or more thick that can be found at each successive exposure studied, because such a bed generally can be recognized even where exposures are poor. The line thus drawn on the map will separate what is nearly all sandstone on one side from what is nearly all shale on the other. The line on the map is a boundary between adjacent rock units. It is not the boundary of a single bed, which would be essentially time-parallel, but a contact that occurs at progressively higher or lower beds as the facies undergo lateral change. Such a boundary is decidedly time-transgressive.

The extreme subtlety of the time-transgressive nature of rock units in the field cannot be overemphasized. In an area of perfect exposures and accessibility, we could follow each individual bed at a facies boundary as it became thinner, less distinct, and finally pinched out entirely amidst the beds of the adjacent facies. In practice this is rarely possible. Ordinarily, we can hope to ascertain variation with respect to some distinctive marker bed that appears to be continuous between good exposures. If, for example, a distinctive marker bed occurred in a shale 20 meters above a gradational contact with a sandstone at one locality and 5 meters above the contact at a second locality some distance away, this would suggest that the top of the sandstone unit rises stratigraphically and becomes younger toward the second locality. The actual contacts may appear identical in both places, and without information to the contrary one might be inclined to consider them time equivalents or nearly so. Their position with respect to the marker bed, however, indicates that the upper portion of the sandstone unit at the second locality is equivalent to the lowermost portion of the shale at the first locality. A facies change here can be recognized only by making a detailed correlation.

Because detailed correlations are usually difficult, the working out of complicated facies changes in a large area is a big job. A well-known facies complex in the United States is the Devonian Catskill Delta in New York State, shown in Fig. 4-2. It is justly famous because it represents one of the finest

FIG. 4-2 Cross section of Devonian System along New York-Pennsylvania border. Names are stage names; heavy lines crossing the facies are time lines. (After D. W. Fisher and others, 1962.)

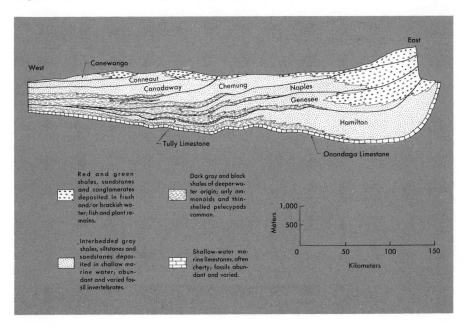

Devonian sedimentary records in the world. It is cited as an example so widely that one takes it for granted as obvious, but it is not. It was one hundred years after the first pioneer investigations in New York State before George H. Chadwick and G. Arthur Cooper worked out the relationships of the varied facies of the Catskill Delta complex. As we understand them today, these facies represent a slow, westward regression of the Devonian sea across New York State.

Similarly grandiose and complicated facies in the Permian System of West Texas were totally baffling when the first geologists came upon them unexpectedly. In the words of one of these pioneers:

> A blackboard drawing or a textbook illustration of a sedimentary facies has quite a different appearance from a sedimentary facies when one encounters it in the field. In the field, facies changes seem baffling and bewildering, especially in an area of new and unknown stratigraphy. It is as if stratigraphy, hitherto subject to natural laws and capable of rational analysis, had suddenly turned lawless and planless.
>
> Those of us who grew up with West Texas Permian geology between 1925 and 1940 learned facies the hard way. I remember vividly, for example, a summer afternoon in the Sierra Diablo in 1928, when we climbed a projecting angle of a canyon wall and saw spread before us the stratigraphy of the Permian rocks of the interior of the range. In the preceding weeks we had been tracing out the stratigraphy of the Permian on the frontal escarpment of the range and had divided the rocks into four or five well-marked lithologic and faunal units of limestone, dolomite, and shale. But now, as we looked into the interior of the range, we saw all our fine units dissolve before our eyes, merging into a monotonous sequence of thin-bedded dolomite that extended as far up the canyon as we could see.*

Ultimately, the facies changes in both the New York Devonian and the West Texas Permian were resolved in the way in which all facies changes must be resolved, through the process of detailed correlation.

A Note on Vertical Exaggeration

The vertical exaggeration obtainable on cross sections is a great advantage in facies analysis, but it can be misleading unless the scale relationships are constantly kept in mind. One becomes so accustomed to cross sections in which the height of the diagram is about the same as the width that it is easy to forget that the thickness of the strata is typically measurable in tens or hundreds of meters, whereas distance is typically measurable in tens or hundreds of kilo-

---

* From Philip B. King, *Sedimentary Facies in Geologic History*, ed. C. R. Longwell, Geol. Soc. America Memoir 39, 1949, p. 165.

meters. Where thickness is exaggerated, angles between time lines and rock boundaries are also exaggerated. On a cross section these angles are easily perceptible. Without thoughtfully analyzing the diagram, one would almost expect to be able to observe the slope of the rock contact in the field, but in the field it can never be perceived.

To illustrate this point, a well-known example of facies change in the transgressive Cambrian strata of the Grand Canyon area is shown in Fig. 4-3(A); the vertical exaggeration of rock thicknesses there is typical. Time lines, based primarily on the two faunal zones and partly on intertonguing rock units, clearly cut across formational boundaries. In Fig. 4-3(B) the vertical exaggeration is relatively small and the angles between time lines and rock boundaries,

FIG. 4-3 Vertical exaggeration in cross sections showing Cambrian facies changes in the Grand Canyon. (A) Vertical exaggeration of 230 times is about that needed to show stratigraphic details. (After E. D. McKee, 1945.) (B) Vertical exaggeration 23 times. (C) At actual scale the thickness is about that of a pencil line.

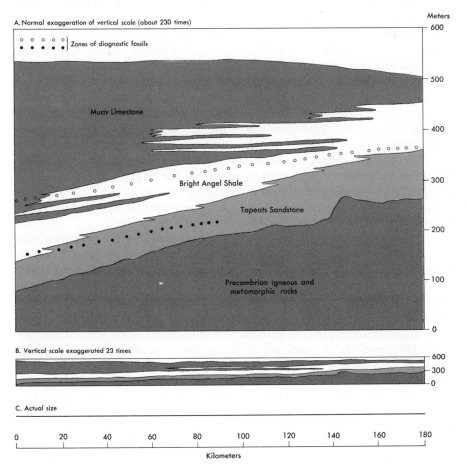

if they could be shown here, would be extremely low. Detail cannot be shown, however, because the vertical scale is insufficient. In the flattening process angles decrease nearly to zero. The same section at true scale would appear only as a line, as in Fig. 4-3(C). Put another way, Fig. 4-3(A) could be corrected to actual scale by increasing the width to about 30 meters.

TIME CORRELATION BY PHYSICAL MEANS

Key Beds

Single beds or laminae that have no bedding planes within them result from single episodes of deposition. To trace an individual stratum, then, is to trace the record of a single, depositional event, and this is the most reliable means of physical correlation. Unfortunately, individual beds can rarely be recognized from outcrop to outcrop with certainty. Particularly in deposits from shallow-water environments, where local currents abound, individual beds cannot be traced more than a few tens of meters. Beds that formed in deeper, quiet water, however, may be widespread, extending over several kilometers. In basins fed by turbidity currents, for example, individual graded beds are sometimes produced over wide areas by an instantaneous influx of sediment, and they are not markedly disturbed by subsequent reworking. On the continental slope off Newfoundland, the Great Banks earthquake of 1929 triggered a turbidity current that distributed sediment over a quarter of a million square kilometers of the deep sea floor in the North Atlantic. Even a bed deposited over only a few tens of square kilometers can be of great correlative value in the stratigraphic record. If such a bed can be recognized throughout its extent by some special peculiarity, it is considered a *key bed*. Recognition is a problem in practice, however, whether one is attempting to trace a bed in a thick sequence of turbidity-current deposits or in some other kind of lithology.

Bentonite beds are especially valuable in physical correlation because they are distinctive kinds of clay that formed by the alteration of volcanic ash in marine environments. Bentonites are only sporadically common in the stratigraphic record, but where they can be traced regionally they make ideal key beds for correlation. They represent simultaneous ash fallout from volcanic eruptions, hence they form in a geologic "instant." Some bentonite beds cover very large areas, easily tens of thousands of square kilometers, which is not surprising in view of the phenomenally large dust clouds produced in large volcanic eruptions. For example, the violent 1815 eruption of Tomboro on the island of Sumbawa, just east of Java, produced dust which caused black darkness for three days across Java 500 kilometers to the west. Significant amounts of dust fell out as far as 1,300 kilometers to the east. Volcanic dust from the 1932 eruption of Quizapú in Chile was borne by westerly winds 1,500 kilometers across South America and well out into the Atlantic Ocean. Where the dust from eruptions such as these falls into sedimentary environments, it produces a widespread instantaneous deposit.

When they are more than about two feet thick, bentonite beds can be recorded on electric logs and they find great utility in subsurface geology where reliable time markers are all too rare. The topmost line connecting the sections in Fig. 4-1 indicates an exceptionally widely distributed bentonite bed in thick Upper Cretaceous shales in eastern Colorado. Figure 4-4 shows an outcrop of this key bed.

FIG. 4-4 Thick bentonite bed in Cretaceous Benton Group of eastern Colorado. This bed is used widely in subsurface correlation.

Position in a Cycle

Events other than volcanic eruptions may influence the stratigraphic record over a wide area and, hence, may likewise be of correlative value. One of these is a *eustatic rise* in sea level, in which sea level is simultaneously elevated worldwide. In the rock record, however, a eustatic rise is difficult to discriminate from tectonic subsidence of the land. Besides, local or regional tectonic movements either add to or subtract from the effects of a eustatic change. Nevertheless, there are notable episodes in the geologic record that are thought to be due largely to eustatic changes. The best-documented of these occurred in the Pleistocene Epoch, when the waxing and waning of icecaps gave rise to worldwide sea-level changes of 100 meters or more.

In a single region the stratigraphic record of a transgressive-regressive cycle contains an inherent time plane, the time of maximum inundation of the sea. If there were no tectonic movements locally within the basin, the strata deposited in the deepest water achieved at any locality should be contemporaneous with strata deposited in the deepest water achieved at any other locality. Correlations based on maximum water depth have been utilized in the thick Cenozoic sedimentary rocks of the Gulf Coast of the United States. In the Gulf region maximum accumulation of sediments occurred approximately at the position of the present coast. Sporadically during the Cenozoic the sea transgressed the land area to the north, producing what appear now as tongues of marine sediments that extend northward into marginal marine and continental facies. Sedimentation was rapid and in critical areas both lithofacies and biofacies change over short distances. The maximum depth of water attained during such transgressive-regressive cycles throughout the region has proved useful in correlating between these diverse facies. In Fig. 4-5, for example, fossil foraminifera, which are known to have lived at different depths in the sea, were used to estimate changing depth of deposition in several subsurface sections. Greatest water depth in each section provides the basis for correlation.

Just after the turn of the twentieth century, large-scale transgressions and regressions of the sea were considered an ideal means of correlation. An eminent American geologist, T. C. Chamberlin, even christened tectonic cycles, which he considered to be their cause, "the ultimate basis of correlation." Chamberlin's

FIG. 4-5 Correlation based on peak transgression during a transgressive-regressive cycle. Influence of the transgression was felt as far north as locality A. Benthonic foraminifera assemblages I–V show that maximum depth of water increases systematically seaward. Barring tectonic complications, strata recording greatest depth at all localities should be time equivalents. (After M. C. Izraelsky, 1949.)

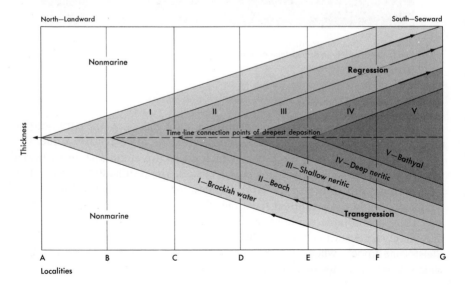

belief stemmed from a popular idea of his day that tectonic activity chiefly affects continents, that this activity is sharply periodic in time rather than continuous, and that the tectonic pulses that affect the Earth's crust are separated by long periods of quiescence. In this view, all continents are uplifted at once, causing simultaneous regression of the seas, which, in turn, produces worldwide unconformities at the same positions in the stratigraphic record. During the following period of tectonic quiescence, land surfaces erode gradually to base-level, sediment is deposited in ocean basins, and the rising seas gradually invade the continents again. Each cycle could be separated into a transgressive, inundative, and regressive phase. Some workers envisioned such cycles as the framework of the geologic systems. The three series of which most systems are composed were supposed to fit the three stages of a cycle. A corollary of the theory held that the system boundaries should ordinarily be represented by unconformities and that systems are, by and large, natural in that each represents a "geologic cycle."

The scheme has not been borne out by the stratigraphic record. Tectonic activity, rather than being periodic, has always been going on somewhere, although at times it affected broader regions than at others. Nevertheless episodes of extraordinarily widespread advances and withdrawals of the sea, like those shown in Fig. 3-11, may have been caused by large-scale changes in the style or rate of sea-floor spreading, which changed the actual water capacity of the oceans and thereby triggered major eustatic sea level changes. These would have produced synchronous regressions or transgressions in widespread coastal regions.

Varves

Varves are couplets of strata produced by seasonal climatic changes. A varved deposit thus consists of a repetitious sequence of annual cycles. Correlations based on varves are useful in special circumstances. Varves are most common in deposits of glacial lakes where summer meltwater brings an influx of silt into the lake, and winter freezing permits the settling out of only the very finest clay particles and organic material from suspension. Inasmuch as the Pleistocene Epoch was a time of ice ages, Pleistocene varved clays, such as that shown in Fig. 4-6, are common.

Individual varves in a varved sequence are not identical. Entire varves or portions of them may be thicker or thinner than average as a result of climatic differences from year to year. A sequence of several varves thus makes a unique record which may be recognizable widely throughout a region. In this way, varves have aided in correlating Late Pleistocene deposits over large areas of northern Europe and other regions as well.

Varves occur not only in glacial lake deposits, but also in certain evaporite deposits and in silled marine basins. The alternating laminae of calcite and anhydrite in the marine Permian Castile Formation of West Texas, for example, have also been interpreted as varves (see Fig. 4-7). The laminations of light

FIG. 4-6 Pleistocene varved clay from Connecticut. Light-colored layers are silty and represent summer deposition. Dark layers are nearly pure clay covered by organic material and represent winter deposition. (Courtesy W. C. Bradley.)

FIG. 4-7 A sample from the Permian Castile Formation of West Texas. Couplets of anhydrite (thick light laminae) and calcite with organic material (thin dark laminae) are believed to represent seasonal climatic changes.

anhydrite are inferred to represent the drier season, and the thin dark laminations of calcite and organic material indicate the more humid season. Thickness of most of these varves is between 0.5 and 2 millimeters. As in glacial lake deposits, variance in thickness provides the means to recognize unique sequences of beds in an otherwise apparently monotonous rock unit. Utilizing such sequences, individual Castile varves have been correlated over distances of 15 kilometers. Apparently their potential is much greater. In the similar Permian Zechstein Formation of Germany, individual calcite-anhydrite couplets have been correlated over distances of 300 kilometers. Varves not only permit remarkably detailed correlation, but in addition they record the number of years that the deposit represents and thus provide a record of the actual duration of particular sedimentary environments. Deposition of the 600 meters of Castile evaporites required about 300,000 years.

Paleomagnetic Correlation

Magnetic iron oxide minerals occur in very small quantities in most sedimentary and igneous rocks. When tiny particles of these minerals settle to the ocean floor or when they cool within a newly formed igneous body, their magnetic polarity aligns itself with that of the Earth's magnetic field at that place and instant of time. Trace amounts of magnetic iron oxide particles thus produce a preferential direction of magnetization in sediments or rocks that contain them. This property is called *remanent magnetism*. Dark igneous rocks show the strongest remanent magnetism, but that of many other kinds of rocks can be measured easily. In simulated rock produced in the laboratory and in geologically young rocks from all parts of the world, the alignment of the remanent magnetism consistently parallels that of the Earth's present field. But in older rocks, the alignment of the remanent magnetism departs from that of the Earth's present field, and, in general, the older the rock, the greater the departure. The direction (declination) of the remanent magnetism, and its angle with the horizon (inclination) may be used to compute the position of the Earth's magnetic poles at the time the rock formed.

Interpretations of paleomagnetic pole positions from remanent magnetism are based on assumptions that should be kept in mind because they may not always be valid. The first assumption is that the Earth's field has always been as it is today—dominantly a dipole, like a bar magnet. The second is that the remanent magnetism of the rock being measured was induced at about the time the rock formed and not at a significantly later time.

Thousands of paleomagnetic measurements from rocks of different ages all over the world indicate that the Earth's magnetic poles have gradually changed position with respect to the continents through geologic time. This phenomenon, called *polar wandering*, provides a potential means of correlation. Computed polar positions for rocks of various ages from an individual continent delineate a curving path across the Earth's surface. Once a polar-wandering

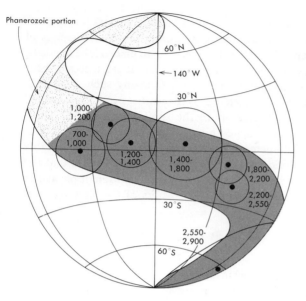

FIG. 4-8 Generalized polar wandering path for Precambrian time established from North America. Dots show mean pole positions for age groupings (in millions of years). Circles are computed for 95 percent confidence level. (After Larson, Reynolds, and Hoblitt, *Geol. Soc. America Bull.*, v. 84, 1973.)

path has been determined for a given continent, then rocks of unknown age could possibly be dated by ascertaining the position of their paleomagnetic poles on the established path.

Preliminary polar-wandering paths have been constructed for most of the continents, but they contain large measurement errors. These show up as a wide scatter of points when the individual paleopole locations are plotted on a map. In part the scatter is caused by errors in the radiometric ages of the rocks whose paleopoles are plotted, but most of the scatter is caused by later recrystallization or alteration of magnetic minerals or by structural deformation of the rocks that contain them. The method's greatest promise is for Precambrian rocks, which, because of their lack of fossils, are difficult to correlate. The precision attainable for Precambrian rocks is suggested by Fig. 4-8, showing a broad belt, determined statistically, within which the North American Precambrian paleopoles probably migrated. This does not show errors in ages of the rocks used in the construction, only errors in paleopole location. If ages are assumed to be precise, the diagram shows a pole migration of about one degree per 10 million years. Hence a 20-degree error in pole position for an unknown North American rock will have a correlation error of about 200 million years. For many Precambrian rocks even this seemingly low degree of precision would provide a helpful refinement in establishing age.

In addition to wandering about, the Earth's north and south magnetic poles have abruptly exchanged positions many times in the geologic past. Each of these paleomagnetic reversals simultaneously affected the field for the entire Earth. Evidence for reversals of the Earth's magnetic polarity that have occurred during the Cretaceous and the Cenozoic has been obtained independently from such diverse sources as terrestrial lava flows and deep-sea sediments. Radiometric dating of the rocks in which the magnetic polarity was measured has proven that each of the polarity reversals was worldwide in scope, and the sequence of reversals is now fairly well established. During the Cretaceous and Cenozoic there have been numerous times when polarity was either dominantly normal, as today, or reversed (Fig. 4-9). These times of generally sustained polarity are called *magnetic polarity epochs*. Within these epochs were apparently short-lived polarity reversals, which have been termed *events*. The mechanism for the abrupt reversals of the Earth's magnetic field is unknown, but it must relate somehow to changes in flow patterns in the liquid iron of the Earth's core, where the main magnetic field originates.

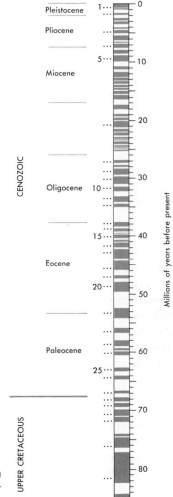

FIG. 4-9   A time scale for magnetic reversals during the last 85 million years. (After Larson and Pitman, *Geol. Soc. America Bull.*, v. 83, 1972.)

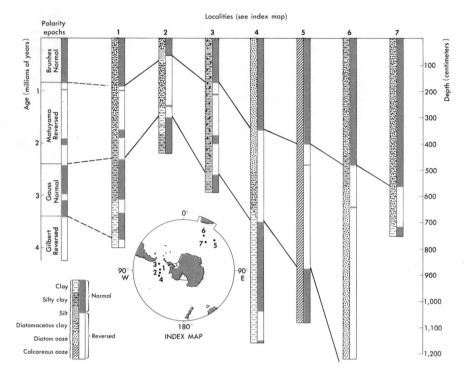

FIG. 4-10 Correlation of seven deep-sea cores from the Antarctic region based on polarity changes in the Earth's magnetic field. (After Opdyke, Glass, Hays, and Foster, 1966, *Science*, v. 154, no. 3747, p. 350. Copyright 1966 by the American Association for the Advancement of Science.)

Figure 4-10 shows the correlation of seven widely spaced cores of deep-sea sediment from the Antarctic region based on their remanent magnetism. The temporal value of the polarity epochs is verified by their constant relationships to time-stratigraphic zones based on fossils. The polarity changes are independent of the markedly different kinds of sediment found in the deep-sea cores. Paleomagnetic reversals thus appear to be an extremely promising tool for correlating and dating deep-sea sediments. Because polarity of the Earth's magnetic field is a global phenomenon, the method has worldwide potential. Cores from ocean-floor sediments from many oceanic regions, and rocks from continental regions as well, have been correlated with those paleomagnetic events shown in Fig. 4-9.

Throughout most of the Cenozoic Era paleomagnetic reversals have occurred with a frequency comparable to that of the last 4.5 million years. However, because of this relatively high frequency their correlation potential prior to a few million years ago quickly falls to zero because there are only two kinds of polarity epochs. After establishing that a rock's magnetism is either normal or reversed, one cannot tell which normal or reversed epoch it represents without an independent means of correlation. If the time span between paleomagnetic reversals is less than the time span that can be resolved by fossil zones or by radiometric dating, then the identity of the paleomagnetic event cannot be determined.

For example, a typical radiometric date for a middle Eocene rock might be determined as 50 $\pm$ 1.5 million years, which places the sample in a 3-million-year time slot. Suppose the rock is normally magnetized. If there is more than one normal epoch in the 3-million-year time span, there would be no way of discriminating which one the rock represents. If, on the other hand, there is but one reversal, say from normal below to reversed above in the time interval in question, then one can refine the correlation to only the lower portion of the interval. In parts of the pre-Cenozoic record, paleomagnetic epochs are very long, which makes reversals of potential value in their particular parts of the geologic column.

LITHOFACIES MAPS

Once correlation has established time-equivalent intervals in stratigraphic sections, the spatial relationships of lithofacies become apparent and can be illustrated on a stratigraphic cross section. If data are sufficient, lithofacies can actually be plotted on a map. Lithofacies maps simply show the areal distribution of rock types of a given age. Their accuracy depends on (1) the accuracy of the correlations used to recognize the time interval and (2) the number of stratigraphic sections available. The areal distribution patterns on facies maps greatly aid environmental interpretations because the sedimentary associations can be compared directly to those accumulating today. Commonly, the thickness of the mapped interval is also shown on lithofacies maps by means of isopach (equal-thickness) lines so that the rock types can be readily related to thicknesses. Location of stratigraphic sections on which the map is based should also be shown so the user can judge the adequacy of the control.

Figure 4-11 shows lithofacies of the Middle Devonian Series in western Canada. The isopach lines show that the Middle Devonian is thickest in the broad central belt of evaporites. It thins eastward and northward into a bedded carbonate facies and westward into a clastic facies. The belt of reefs on the north edge of the evaporite basin appears to have separated the hypersaline water to the south from the more normal marine water in which the platform carbonates were deposited to the north. The belt of clastic rocks on the west was apparently derived from adjacent sources, and thus one can infer a nearby exposed land area immediately to the west.

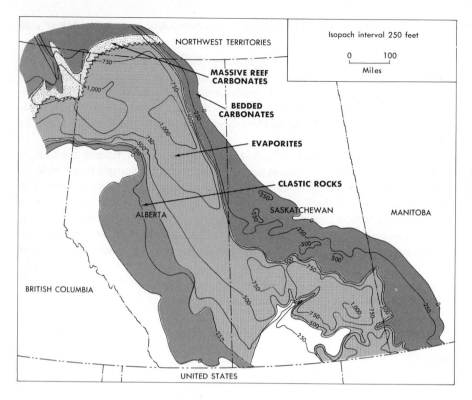

FIG. 4-11 Lithofacies and thickness map of the Middle Devonian Series of the Canadian prairie provinces. (After R. G. McCrossan and others, 1964.)

PALEOGEOGRAPHIC MAPS

From inferences based on the facies, it is possible to make an environmental map showing marine, marginal, and continental environments. Maps that show the gross environmental geography of a region are called *paleogeographic maps*. Figure 4-12 is a paleogeographic map for a small portion of Mississippian time in the east central United States. Here deltas are shown building westward and southward into the shallow midcontinent sea from land areas on the north and east. Paleogeographic maps ideally represent instants in time, and a series of them is required to convey the total sequence of events that transpired during a depositional episode. Figure 4-12 is one of a series of maps that document the evolving Mississippian paleogeography in this region.

Valid correlations are vital in interpreting the geologic history of an area correctly. A backward order of events may easily result when boundaries of lithologic units are hastily interpreted as time planes. Figure 4-13 shows a sea transgressing a former land area. Close to shore, sand is deposited; further from shore, mud is deposited; and far out to sea, beyond the reach of detrital material, calcareous sediment is deposited.

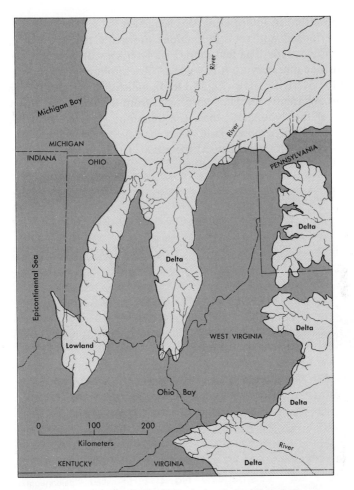

FIG. 4-12 Paleogeographic map of the Early Mississippian sea (shaded area) and adjacent marginal and continental environments in the east-central United States. (After Pepper, DeWitt, and Demarest, 1954.)

FIG. 4-13 Block diagrams showing successive shoreward positions of a transgressing sea in which sand is deposited nearshore, mud offshore, and calcareous sediment still further offshore. The top surface of the two blocks represents paleogeographic maps for time A and a later time B.

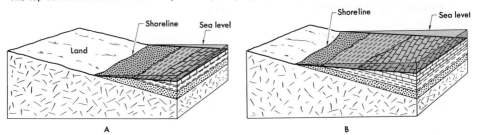

Suppose that in interpreting this sequence the contemporaneity of these units is overlooked and rock units alone are made the basis for a series of paleogeographic maps. The first map (Fig. 4-14(A)) will show the extent of the basal sandstone unit because it is "oldest." Then follows a map showing distribution of the overlying shale (Fig. 4-14(B)), indicating that during "shale time" the sea had apparently retreated somewhat from the land area. Finally, during deposition of the "youngest" unit, the limestone (Fig. 4-14(C)), we find the maximum extent of exposed land. Here paleogeographic mapping based on formation-tracing alone has caused the record left by a transgressing sea to be incorrectly interpreted as a regression.

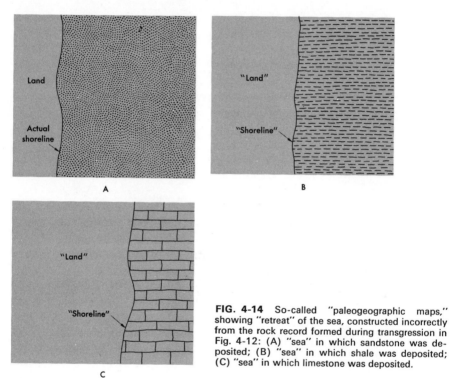

FIG. 4-14   So-called "paleogeographic maps," showing "retreat" of the sea, constructed incorrectly from the rock record formed during transgression in Fig. 4-12: (A) "sea" in which sandstone was deposited; (B) "sea" in which shale was deposited; (C) "sea" in which limestone was deposited.

CHANGING POSITIONS OF THE CONTINENTS

A good paleogeographic map for a given time in the geologic past indicates a comprehensive understanding of geologic history. Excellent maps, like that of the Mississippian of the east central United States in Fig. 4-12, indeed reveal a sound historical knowledge of large areas on the continents. But what about relationships between continents? If we expanded Fig. 4-12 to show the region east of the Appalachians, would a Mississipian Atlantic Ocean be about where it is today, or would the northwest coast of Africa adjoin the Atlantic seaboard? A global map that embodies the concept of continental drift looks a great deal different from one that does not. The question of whether or not the continents always occupied their present sites on the Earth's surface was warmly debated for many decades, and in the past few years it has been answered decisively, chiefly by the evidence from paleomagnetism, from radiometric dating, and from the results of deep-sea drilling. The continents almost certainly have moved large distances, relative to one another, since the beginning of the Mesozoic era. However, it is not yet clear how extensively they moved prior to this time.

The stimulus for the idea that the continents were once joined was the striking jigsaw puzzle fit of shorelines on opposite sides of the Atlantic. This remarkable fit, and that of the continents around the Indian Ocean, are still highly appealing lines of evidence for continental separation. The best fit is obtained, not at the present coastlines, but at a depth of 900 meters, about halfway down the continental slopes. There the continents in the northern and southern hemispheres may be matched with only small gaps and overlaps that add up to an average misfit of just 130 kilometers (see Fig. 4-15).

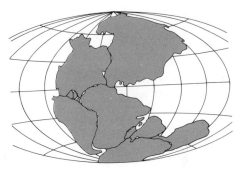

FIG. 4-15 A reconstruction of the probable relationships of the continents in the late Paleozoic, prior to their breakup, as envisioned by Dietz and Holden.

Stratigraphic Evidence

Strata ranging from Carboniferous to Jurassic in age are remarkably alike in all the Southern Hemisphere continents and islands. These strata are largely nonmarine in origin, and the similar geologic history they indicate supports the idea that, at the time they formed, all of the Southern Hemisphere lands were part of a single huge continent, called *Gondwanaland*. A prominent boulder-bearing formation at the base of this stratigraphic sequence in South America, the Falkland Islands, South Africa, Madagascar, India, Australia, and Antarctica has been interpreted as a *tillite*; that is, as lithified glacial till. These deposits provide excellent evidence that huge continental ice sheets occupied vast regions of the Southern Hemisphere continents in the Carboniferous and Permian Periods.

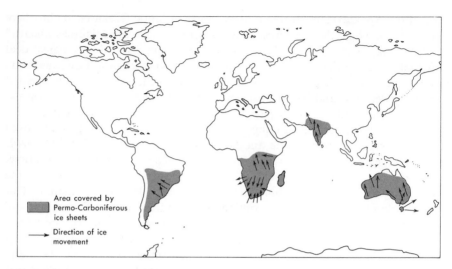

FIG. 4-16 Ice flow directions away from present ocean basins and tropical occurrences of Permo-Carboniferous glacial deposits is strong evidence for continental drift.

The late Paleozoic tillites are distributed from Antarctica as far north as the tropics in India, Africa and South America (Fig. 4-16). Because the occurrence of continental ice sheets in the tropics is most unlikely, these tillites are strong evidence that southern continents were probably not always in the same positions with respect to the Earth's equator. Regardless of where the south pole is placed within the present arrangement of land masses, however, some of these ancient glaciated areas still lie within 10 degrees of the equator. This suggests that during the Permian and Carboniferous Periods, the Southern Hemisphere continents were clustered closely about a Late Paleozoic south pole, the expected geographic center of the glaciation.

Supplementary evidence that Southern Hemisphere continents were joined comes from eastern Brazil where, judging by the grooves in the underlying bedrock, the ice flowed dominantly westward from what is now the Atlantic Ocean. Moreover, the tillite itself contains boulders of distinctive quartzite, unlike any known from South America but identical to rocks that crop out in western Africa. If Africa was indeed the source area for the Brazilian continental ice sheets, certainly the South Atlantic Ocean could not yet have existed.

Stratigraphic evidence of a completely different kind comes from drilling in the deep ocean basins. At many sites in the Atlantic, for example, cores have penetrated the entire sedimentary section and bottomed in the underlying basaltic crust. At sites progressively nearer the Mid-Atlantic Ridge, the oldest sediments—immediately overlying the basaltic crust—are found to be progressively younger. This is considered excellent evidence that the ocean floor itself decreases in age toward the ridge.

Paleontologic Evidence

Permian strata above the basal tillite everywhere in the Southern Hemisphere contain a unique assemblage of plants called the *Glossopteris* flora, which is unknown in the Northern Hemisphere. Southern Hemisphere lands were thus a unique Permian floral province, and advocates of continental drift have long pointed to this as evidence that the continents were connected.

In both Africa and South America a small Permian reptile, *Mesosaurus*, occurs in deposits of freshwater lakes immediately above the basal tillite. Apparently *Mesosaurus*, or his immediate ancestors, was able to migrate between Africa and South America. Inasmuch as *Mesosaurus* is nonmarine in habit, the Atlantic Ocean surely would have acted as a barrier unless, of course, it was not there.

Structural Evidence

Many of the structural features on the west coast of Africa seem to have counterparts at corresponding positions on the east coast of South America. Although much of the structure of the fringing continental shelves is unknown, the structural compatibility is about as good as could be expected if Africa and South America were once joined. Similarly, great structural lineaments of the eastern United States and Canada match up remarkably well with similar features in Scotland and Scandinavia.

The foregoing is only a sample of the kinds of geologic evidence in support of continental drift that have been known for many decades. Until the late 1960's this evidence failed to convince the many adherents to the idea of continental permanency, who were strongly influenced by the physicist's admonition that the Earth's crust was far too rigid to have permitted migration of the continents. Then persuasive evidence from radiometric dating, and especially from paleomagnetism, began to accumulate rapidly and this convinced the scientific world that continents have moved huge distances and that, even today, large segments of the Earth's crust are moving slowly in different directions.

Evidence from Radiometric Ages

Striking support that Africa and South America were once joined comes from the match of pairs of Precambrian orogenic provinces of different ages in Brazil and West Africa. The West African Precambrian Shield is divided into two distinct orogenic provinces similar to those that have been recognized on the Canadian Shield (see Fig. 4-17). Rocks of the Eburnean Province on the west consistently yield dates of about 2,000 million years. Rocks of the Pan-African Province on the east consistently yield dates of about 550 million years. When South America and Africa are joined in the postulated predrift configuration shown in Fig. 4-14, counterparts of the West African Shield might be

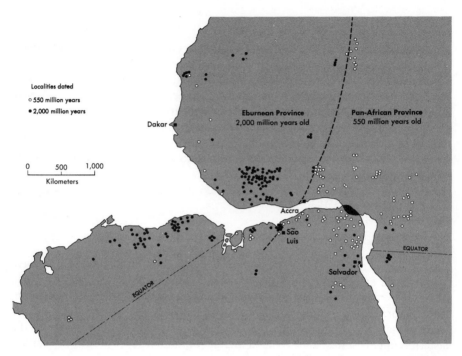

FIG. 4-17 Boundary between 2,000-million-year Eburnean age province (solid circles) and 550-million-year Pan-African age province (open circles) appears to extend from West Africa directly into a predicted location in Brazil when Africa and South America are fitted together according to the reconstruction of Bullard and others (see Fig. 4-14). (After P. M. Hurley and others, 1967, Science, v. 157, no. 3788, p. 496. Copyright 1967 by the American Association for the Advancement of Science.)

expected on the adjacent Brazilian Shield, and the sharp, southwest-trending boundary between them should extend onto Brazil near the city of Sao Luis. These facts, predicted by the drift theory, were amply confirmed by a group of American and Brazilian geologists who collaborated in dating a large area of the Brazilian Shield adjacent to the Atlantic seacoast. The precise match of the Precambrian provinces on opposite sides of the South Atlantic strongly supports the theory of continental drift because the theory actually predicted the results in advance.

Evidence from Paleomagnetism

If two continents maintained the same relative positions, then the traces of their polar-wandering paths for rocks of the same age should coincide. But, in fact, no two continents share the same polar-wandering paths. This must mean that the continents have moved relative to one another. Figure 4-18 shows that when the calculated Paleozoic polar-wandering paths for Africa and South America are superposed, Africa and South America move together in the same positions suggested by their jigsaw reassembly in Fig. 4-15. The

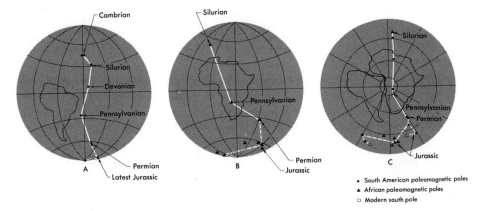

FIG. 4-18 Paleozoic and Early Mesozoic polar-wandering paths indicated (A) from South American rocks and (B) from African rocks. When the Paleozoic portions of the curves are brought together as in (C), Africa and South America join. (After K. M. Creer, 1965.)

Jurassic poles, however, do not coincide, suggesting that by that time the continents had begun moving apart. The Paleozoic portions of polar-wandering paths for the other continents also become superposed when the continents are arranged as shown in Fig. 4-15. The paths diverge in the Mesozoic, but not at the same rate. For example, the North Atlantic began to open in the Triassic, but the South Atlantic did not open significantly until after the late Jurassic, the time represented by the paleogeographic map in Fig. 4-19.

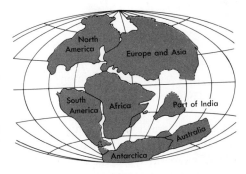

FIG. 4-19 Paleogeography of the continents and deep ocean basins at the end of the Jurassic Period 136 million years ago, as envisioned by Dietz and Holden. The seaway between Africa and Europe is closing. The North Atlantic and Indian Oceans have opened considerably. The South Atlantic has been initiated along a rift. Australia has not yet separated from Antarctica. Shallow seas on the continents are not shown.

If the Atlantic Ocean is enlarging, new oceanic crust must be forming within it, and present evidence indicates that the locus of crust formation is along the Mid-Atlantic Ridge. The Mid-Atlantic Ridge stands more than 1,000 meters above the deep ocean areas on either side, and it exhibits high heat flow, local vulcanism, and shallow earthquake activity. In addition, the crest of the ridge consists of linear fault troughs, which are interpreted as tensional features created by the separation of the crustal blocks on either side. These characteristics suggest that the ridge is a huge pull-apart zone where new crustal

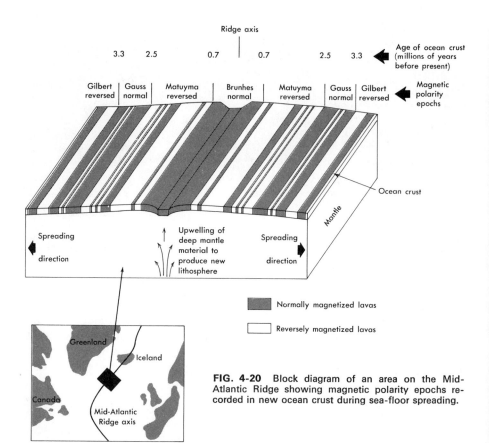

FIG. 4-20 Block diagram of an area on the Mid-Atlantic Ridge showing magnetic polarity epochs recorded in new ocean crust during sea-floor spreading.

material continually wells up from the mantle below to fill the gap left by the separating blocks (see Fig. 4-20).

The hypothesis of the Mid-Atlantic Ridge as a pull-apart zone is strongly supported by subtle differences in the magnetic field on the flanks of the ridge, which are detectable by a highly sensitive instrument called a *magnetometer*. These variations in the magnetic field, which are called *magnetic anomalies*, record a repetitive pattern of stripes that parallel the ridge for hundreds of miles. Moreover, the pattern of anomalies east of the ridge is a mirror image of that west of the ridge (see Fig. 4-20). These linear anomalies are interpreted as alternating bands of normal and reversed polarity in the remanent magnetism of the ocean-floor basalts. In this view, new crustal material cools as it comes near the surface at the crest of oceanic ridges, recording the magnetic field for that time, and then it separates and spreads away from the ridge in both directions carrying its paleomagnetic record with it. The "clincher" in this argument is that one can actually identify, on both flanks of the ridge, the magnetic polarity epochs of the last 4.5 million years (compare Fig. 4-9 with Fig. 4-20).

Additional linear magnetic anomalies farther from the ridge crest record still older reversals.

Plate Tectonics and Paleogeography

The Mid-Atlantic Ridge is only a portion of the 60,000-kilometer oceanic ridge-rise system, which is bordered, throughout its extent, by similar magnetic anomalies. Known ages of key anomalies and exact distances from their sites of origin at ridge crest indicate that present rates of "sea-floor spreading" vary from place to place but are generally between 1 and 10 centimeters per year. Thus continental drift is now viewed as a mere by-product of the motion of huge slabs, or *plates*, which make up the outer shell of the Earth. The *lithosphere*, of which these plates are made, constitutes the outer 100 kilometers of the Earth, and this includes the uppermost part of the mantle as well as the overlying oceanic and continental crust. The lithosphere rests upon a nonrigid, partly molten zone in the mantle, which is called the *asthenosphere.*

If vast quantities of new lithosphere material are being generated at pull-apart zones, then either the Earth is expanding or equivalent quantities of lithosphere are being consumed somewhere else. Most workers feel that expansion on the scale required is extremely unlikely. They believe that old lithosphere is sinking deep into the mantle along linear zones of plate convergence, which are called *subduction zones.* Subduction zones are believed to be located below the oceanic trench systems, chiefly in the Pacific. These are sites of the deepest-known earthquakes and the largest-known negative gravity anomalies, which suggest that relatively light lithospheric material is being forcibly dragged down, or "subducted," into relatively heavy, deep mantle material.

In addition to pull-apart zones and subduction zones, there are numerous boundaries where lithospheric plates are not being created or destroyed but are simply slipping sideways, one past the other. Together these three kinds of boundaries define 7 huge plates and 20-odd small plates that completely encompass the Earth's surface. Each plate moves independently as a coherent unit. This entire scheme of so-called plate tectonics has been widely embraced only since the late 1960's. Although several important problems have not yet been explained, the general scheme as outlined above is supported by convincing evidence. Clearly the positions of ancient lithospheric plates must be reckoned with in order to have any meaningful large-scale reconstructions of ancient geography.

five

biostratigraphy

The term "biostratigraphy" refers to the dating and correlating of rocks by means of fossils. Fossils have left a record of life on this planet that stretches back more than 3 billion years. This record testifies that all organisms have undergone constant evolutionary change. In the Precambrian rocks this record is meager, but in the Phanerozoic rocks it is elaborate indeed. Geologists have arranged the myriad of species from these richly fossiliferous strata in their proper sequence and have thus established a time framework for the Phanerozoic.

However, not all changes in a vertical succession of fossil-bearing strata are caused by organic evolution. Particular plants and animals have always been adapted to particular environments. Thus, fossils contain much information about the ever-changing conditions on the Earth's surface. In this role they add significantly to the environmental record that is readable from rock types alone.

The Earth's record of life is thus a result of environmental influences superposed on the basic evolutionary changes. And while similar environments occurring at widely different times in the geologic past have commonly produced similar sedimentary rocks, similar environments widely spaced in time have never yielded the same faunas and floras. Faunas and floras have continually changed as new species appeared and the old ones became extinct. Thus, the total organic spectrum at any given time in the geologic past was unique. Properly interpreted, fossils provide the most accurate means of determining geologic ages of Phanerozoic rocks.

GEOGRAPHIC DIFFERENCES IN ORGANISMS

Species of animals and plants today are distributed in complex local patterns governed by environments. Most species are also confined to large geographic regions bounded by barriers to dispersal. *Ecology* deals with the relationship of organisms to their environments. *Biogeography* deals with the broad distibution of animals and plants on the Earth's surface. *Paleoecology* and *paleobiogeography* are the sciences of ecology and biogeography applied to fossils.

Biofacies

Within a given region today certain organisms occur together repeatedly. Each of these assemblages of species favors a particular habitat—an area that may be small or large—having specific environmental conditions. Adjacent areas with differing environmental conditions support different assemblages. Figure 5-1, for instance, shows the distribution of three different assemblages of benthonic foraminifera in the northern Gulf of Mexico. The *Miliammina* assemblage occurs near the shore in waters of low salinity. The *Ammotium* assemblage occurs in water of moderate salinity, chiefly in the Mississippi Sound. The open

FIG. 5-1 Present-day distribution of benthonic foraminifera in a portion of the northern Gulf of Mexico. (After F. B. Phleger, 1954.)

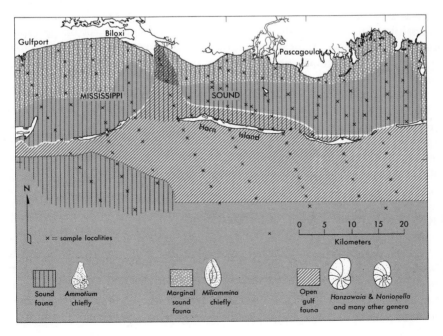

sea south of Horn Island supports a varied benthonic foraminiferal assemblage including *Hanzawaia* and *Nonionella* among others. Farther seaward a sequence of similarly distinct bottom assemblages (not shown) occurs with increasing water depth. Presently accumulating sediments that contain the organic remains of these diverse assemblages will appear as distinct foraminiferal biofacies in the future sedimentary record.

Each species of animal or plant produces more young than can survive in its living space, and thus each tends to expand into every area where environments permit. Highly mobile animals like birds have obvious means of rapid dispersal. But even sessile organisms like land plants produce seeds that may be borne widely by animals or by wind, and many sessile benthonic marine animals have free-swimming larvae that may be carried widely by ocean currents. When a new species is introduced into a new region, it migrates rapidly into favorable environments throughout that region.

Biogeographic Provinces

Few species of plants or animals are distributed throughout the entire world. Most occur only within the confines of large regions called biogeographic provinces. Biogeographic provinces are separated by physical or climatic barriers to the dispersal of organisms. Land areas are barriers for marine organisms and open water is a barrier to the dispersal of most land animals and plants. For marine organisms adapted only to shallow water, a deep ocean basin poses as great a barrier as a land mass. Climatic barriers are effective for many other groups of animals and plants. Barriers that are impenetrable for one group of organisms may be overcome by another group. Some species, for example, have much wider temperature tolerances than others. Geographic limits of swimming marine organisms likewise differ from those of bottom dwellers.

From the ecologic and geographic differences in living species today, we infer that similar differences must have existed in the past. This is verified by the geologic record. Phanerozoic strata contain a geographic variety of fossil organisms comparable to that of modern biologic communities and provinces. The result in stratigraphic terms is an array of biofacies in the rocks that is fully as complex as the lithofacies discussed earlier.

CHANGES IN ORGANISMS THROUGH TIME

Each species today is delicately adjusted to a set of specific environmental circumstances, sometimes termed an "ecologic niche." If conditions remained absolutely fixed, species might never change. Most genetic mutations could only interfere with their delicate ecologic adaptations, and mutating individuals would be selected against. But in nature conditions are nowhere perfectly static for long, and environmental change is the rule, not the exception. Since their first appearance on this planet, animals and plants have accommodated to these changes by slowly evolving through the natural selection of those random mutations that happened to bring the species into better adjustment with shifting environmental conditions. The evolutionary process has been one-directional and nonreversible.

Rates of evolution vary greatly. Some groups of organisms have evolved very little over long stretches of time, while others have evolved at tremendously fast rates, comparatively speaking, and have left a record of striking change in the rocks. Such groups, if they became widely dispersed, provide valuable short-ranged species for time stratigraphy. Paleontologists very early became aware that every geologic period has its characteristic groups of short-lived species or genera and that the stratigraphic value of biologic groups varies greatly from system to system. Figure 5-2 illustrates the time-stratigraphic utility of the major groups of marine invertebrates for various times in the Phanerozoic.

FIG. 5-2 Relative time value of major groups of marine invertebrates during the Phanerozoic. (After C. Teichert, 1958.)

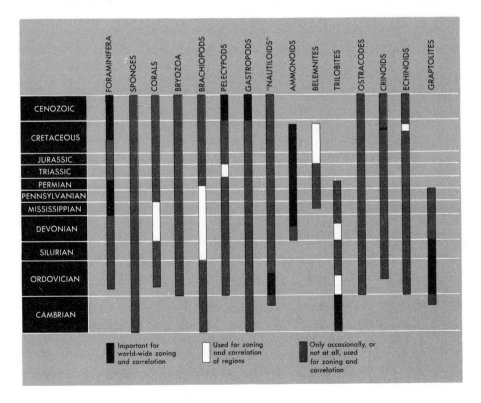

COMBINED EFFECTS OF
GEOGRAPHIC DIFFERENCES
AND TIME CHANGES IN ORGANISMS

The complications of biofacies and biogeography seem to be bothersome impediments when one is attempting to work out time-stratigraphic significance of fossiliferous stratigraphic successions. In a broader sense, however, we are fortunate to have organic records of ancient environments and provinces, just as we are fortunate to have the time utility provided by evolution. Both the environmental and temporal records are important for interpretation of geologic history.

If faunas and floras of the geologic past had not been confined to specific environments and to biogeographic provinces as they are today, and if all evolutionary changes had been simultaneous and worldwide, then even small evolutionary changes would be reflected everywhere in sedimentary rock sequences. We would be able to correlate sedimentary rocks everywhere with great confidence, because all over the world at a given time they would contain identical fossil species. On the other hand, fossils would provide no aid in detecting different ancient environments. In the absence of environmental and provincial distinctions, diversity would be monotonously low, and if only one biotic assemblage existed over the Earth's surface at any one time there would be no faunal basis for interpreting paleogeography or paleoclimates. The world in geologic history would indeed be biologically so monotonous, so unlike that of the present day, that there would be little of interpretive value to be gained from all the precise correlations that the biologically uniform circumstances would permit.

If on the other hand, organic evolution had never occurred and if living things throughout all past time were the same as they are today, all changes in fossil assemblages in sedimentary rock successions would be due either to local changes in environmental conditions or to the breakdown of barriers between biogeographic provinces. Successive fossil assemblages would potentially differ as dramatically as elephants differ from sea shells, but as environments recurred, they would bring with them identical organisms time and time again. On a local scale we would be able to interpret ancient environments with great confidence because fossil assemblages would contain exactly the same species as modern organic communities. However, in our hypothesized absence of evolutionary changes with time, we would not be able to correlate our elaborate local chronologies.

Fortunately, the real situation lies somewhere between the extremes outlined above. This means that with discrimination fossils permit both environmental and temporal interpretations. The effort required to untangle the effects of the two causes is the key to correct reconstructions of geologic history.

THE BASIS FOR BIOSTRATIGRAPHIC ZONATION

No plant or animal has existed for all geologic time. Each species evolved from some ancestor and so had a beginning. If it is extinct today, it also had an ending. Each extinct species therefore must divide geologic time into three parts: the time before it evolved, the time during which it existed, and the time since it became extinct. Any rocks that contain the species must have been deposited within the time during which the species existed. If the total time during which the species existed is short, then its presence in the rocks provides precise placement in the geologic time scale. But if the total time during which the species existed is great, then sedimentary rocks in different regions might well contain the species and yet differ widely in age. The total time range of some species is much shorter than that of other species, and short-ranging species are potentially of superior time value. If such short-ranging species occur widely and are readily recognizable they are sometimes referred to as "guide fossils" or "index fossils."

The strata that actually contain the fossil remains of a species constitute that species' *range zone*. The range zone at most localities represents only part of the total time during which the species existed because the species did not appear everywhere at exactly the same time and did not expire everywhere at exactly the same time. The total time during which a species existed must generally be inferred from the results of extensive collecting over wide regions in order to find the very oldest and youngest strata in which the species occurs. The total time range of a fossil species is always subject to extension upward or downward by new fossil finds. Yet, as more and more observations are made, the probability becomes small that ranges of individual species will have to be revised significantly.

At a locality where a species' range zone is overlain by strata containing its direct descendents and underlain by strata containing its immediate ancestors, the range zone very likely represents the species' total range in time. This situation is rare. Far more commonly a species appears in the fossil record fully developed with no indication in the strata below of its ancestors. We may infer that the range zone of such an abruptly appearing species represents only a part of its total range in time. Yet such a species may be of great biostratigraphic value in certain regions, particularly if its first appearance represents the breakdown of a barrier and its sudden migration into a new biogeographic province. However, if its first appearance in an area is due merely to the shifting of a local environment in a province where the species has existed for a long time then the species is probably an environmentally sensitive *facies fossil*, and of little value in time correlation. The proper interpretation of a species' range zone boundaries is thus important and must generally be based on the evidence from more than one locality.

If the species first appears at a lithologic change (for example, at the base of a sandstone unit whereas it does not occur in an underlying shale), then the environmental change responsible for the new lithology was probably favorable for the species and the appearance is of no biostratigraphic significance. But if the regional evidence indicates that the abrupt appearance is synchronous, or as nearly so as we can tell, then it may represent the species' sudden dispersal throughout the province some time after it evolved elsewhere. The appearance thus represents the surmounting of a barrier by the species, and its immediate migration beyond its former confines. This migration event provides a time-stratigraphic marker that is as useful as a first appearance due to evolution. The biostratigraphic utility of the abruptly appearing species is simply based on that part of its time range during which it existed within the biogeographic province in question.

The top of a species' range zone is commonly abrupt also. If the last appearance accompanies a lithologic change it may signify a mere environmental shift. But if regional evidence indicates that the disappearance represents an extinction event, then it has biostratigraphic value.

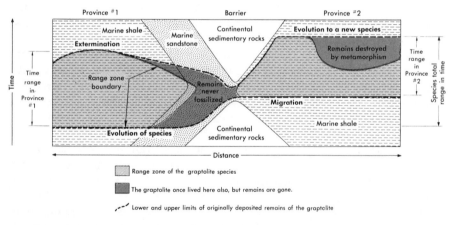

FIG. 5-3 Relationship of a hypothetical graptolite's range zone to its total range in time in two biogeographic provinces. (Modified from H. D. Hedberg, 1965.)

In summary, the correct interpretation of the fossil record requires the discrimination of local environmental effects from time-significant events. Such time-stratigraphic events may be grouped into three categories: evolutionary events, migration events, and extinction events. Figure 5-3 shows an example of these for a graptolite species, and it shows the relation of the rocks that contain the species (that is, its range zone) to the total time during which the species existed and to its time range in two different faunal provinces.

INTERPRETING THE RECORD

Three examples will help to illustrate how range zones, which are the fundamental biostratigraphic units, are recognized and interpreted in the field.

Discriminating Time-Significant Events in Range Zones

This example focuses particularly on the kind of evidence one actually looks for in order to distinguish the effects of time-significant events from the local effects of shifting environments. Refer for the following discussion to Fig. 5-4. The middle of this diagram shows three widely spaced stratigraphic sections, A, B, and C. At each locality a black shale, calcareous in its lower portion, is overlain conformably by limestone. Reconnaissance examination of the three sections reveals that the limestone and the lower calcareous part of the shale contain "shelly" faunas consisting chiefly of brachiopods, and that the upper, noncalcareous part of the shale contains only graptolites. Immediately this distribution suggests a threefold division into gross biostratigraphic units. The strata that contain these diverse assemblages may be called *assemblage zones*. The recognition of these zones, shown in the top part of Fig. 5-4, simply takes into account the overall aspect of the different successive faunas without regard to ranges of individual species. These assemblage zones reflect changing environments, and thus they are of interest in interpreting environmental history. However, as they are followed laterally they will surely cross time boundaries. In order to work out the time stratigraphy we must seek out synchronous events on the species level. For further analysis we will select only a few species that show particular promise as time indicators.

Careful collecting from the shale beds throughout assemblage zone 2 at locality A reveals three successive graptolite species (2, 3, and 4), which are selected for interpretation. These are illustrated next to the column for locality A. Careful collecting from assemblage zones 1 and 3 reveals several distinctive brachiopod species. One from each zone is also selected for interpretation, and these are labeled 1 and 5 adjacent to the column for locality A. The vertical ranges of species 1 through 5 are carefully determined in the field at locality A and at localities B and C as well. The actual vertical limits of each of these species constitutes its range zone. The five range zones for each locality are plotted next to the stratigraphic columns.

Study of the 3 graptolite species from assemblage zone 2 at locality A reveals gradual vertical changes between species 2, 3, and 4, which we infer to be a phyletic lineage. That is, graptolite species 3 is ancestral to 4 and descended from 2. The range zone of graptolite 3 may thus be interpreted as a time-stratigraphic zone because its boundaries are inferred to be unique evolutionary events. They are the bases for inferred time lines W and X in the lower portion

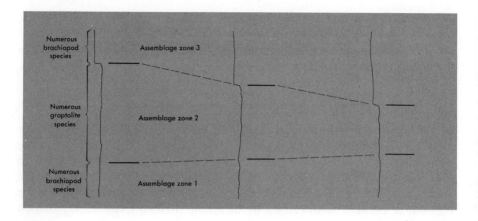

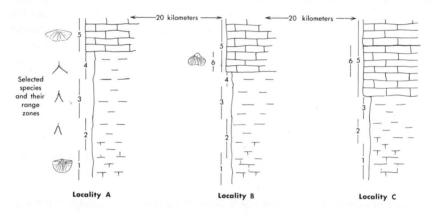

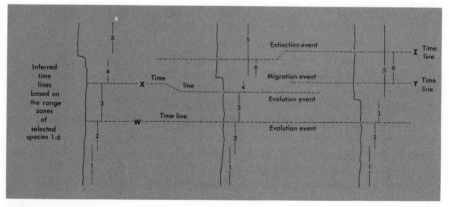

FIG. 5-4 Assemblage zones, selected range zones, and inferred time lines in a time-stratigraphic interval at three localities.

of Fig. 5-4. The most obvious faunal changes in the section—those between the graptolite-bearing and the brachiopod-bearing strata—are interpreted as wholly environmental in origin and hence probably of no time value. Obviously these changes cannot be evolutionary; brachiopods did not evolve into graptolites and then graptolites back into brachiopods.

Locality B seems to bear out these interpretations. Gradations between graptolite species 2, 3, and 4 occur here in the same way as at locality A, verifying the inference that they constitute a phyletic lineage. However, species 4 appears only for a short stratigraphic interval before it is truncated by the brachiopod-bearing strata of assemblage zone 3. This suggests that the base of assemblage zone 3 is older at locality B, which is to be expected if assemblage zones 2 and 3 are environmentally controlled facies. This relationship is verified at locality C where graptolite species 4 has been replaced entirely by the descending limestone facies, which here truncates the upper part of the range zone of graptolite species 3. There is no way to extend time line X into the stratigraphic section at locality C with precision because the evolutionary appearance of species 3, on which it is based, is not recorded there. However, its approximate position is apparent.

At locality B, the lower portion of the limestone unit contains a distinctive new brachiopod (species 6), which is clearly not ancestral to any of the species in the upper part of the limestone. Its lowest occurrence is at the limestone-shale contact, like that of species 5, and thus it is probably environmentally controlled. At locality C, however, the lower limit of species 6 lies well within the limestone unit, suggesting that its introduction at this locality might possibly be due to something other than local environmental conditions; yet, just as it has no overlying descendants, species 6 has no underlying ancestors. Apparently species 6 first evolved somewhere else and was later introduced into our area by either (1) shifting local environments or (2) the breaking of a distant barrier between biogeographic provinces.

Inasmuch as biofacies and lithofacies boundaries do not necessarily coincide, shifting local environments cannot be ruled out simply because the limestone appears identical below and above the first appearance of species 6. In searching for additional evidence, we learn that the phyletic lineage of brachiopods to which species 6 belongs is well represented in strata of this age in another part of the world. We also find that throughout our own biogeographic province species 6 commonly first appears alone and not in conjunction with other species. From these facts we infer that the level at which species 6 first appears at locality C is not due merely to shifting local environments, but instead that it is probably due to migration from another province following the breakdown of a distant faunal barrier. Immediately upon immigrating, species 6 dispersed rapidly throughout our province wherever environments were suitable. Although only part of the total time range of species 6 is represented in our province, the "migration event" with which species 6 is introduced (see the bottom diagram in Fig. 5-4) is considered as good a time marker as an evolutionary event, and it provides the basis for time line Y; obviously, however, it must be interpreted with caution. Locally, as at locality B, the first appearance may be environmentally controlled.

The simple disappearance upward of species 6 is less certainly time-significant. Inasmuch as it left no descendants in our province, there is no evolutionary event to mark its termination. If widespread environmental change caused its disappearance, then the top of its range may mark a nearly synchronous event. However, if local changes caused its disappearance, then even within one province it might persist in some areas long after being removed from others. If species 6 inhabited more than one province, the probability is low that it disappeared at exactly the same time from all of them. In this case its disappearance from each province is only an extermination and the final extermination its extinction. We find that species 6 seems to disappear everywhere in our province at about the same level as it does at localities B and C. Hence we may consider the top of its range zone at these localities a time-significant extermination or extinction, which is labeled on the bottom diagram of Fig. 5-4 simply as "extinction event."

In summary, this example has shown how the three principal kinds of time-significant events in biostratigraphy typically appear in the stratigraphic record. The primary ones are provided by evolutionary changes along phyletic lineages. Migrations and extinctions provide secondary time-significant events. These may be as valuable as evolutionary events, particularly when they involve rapidly evolving groups.

The Mapping of Range Zones

The Upper Cretaceous Pierre Shale, a thick, widespread formation in the Great Plains of the western United States, has had its key range zones mapped throughout large areas. This is unusual. The common practice is to map only rock units, and to express their component fossil range zones only in plotted stratigraphic columns such as those shown in Fig. 5-4. Hence the Pierre Shale provides rare insight into the extent and continuity of range zones. Figure 5-5 is taken from a portion of a larger map showing the areal distribution of range zones of 15 species of the ammonite genera *Baculites*, *Didymoceras*, and *Exiteloceras* in a region where the Pierre is about 2,500 meters thick. The map was constructed from faunal occurrences at several hundred localities in the field. In addition to the 15 range zones the map also shows two sandy members (Kph and Kplr) near the middle and another (Kpt) at the top of the Pierre. A striking characteristic of the map patterns of the range zones is the way in which they illustrate the geologic structure in areas of thick homogeneous lithology where they otherwise would not show.

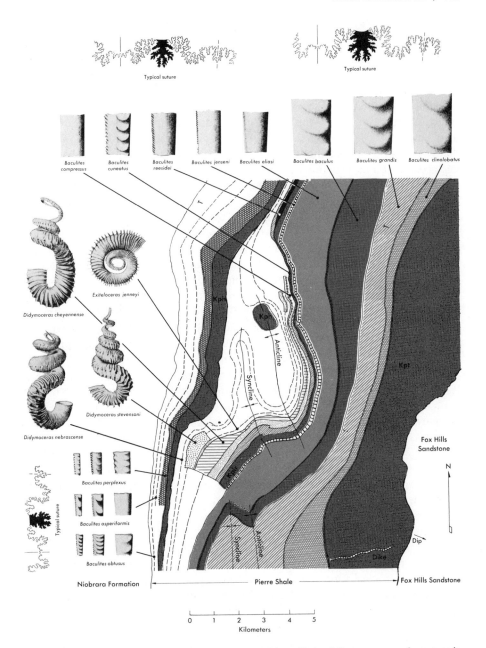

FIG. 5-5 Map of range zones in the 8,000-foot thick Pierre Shale of Cretaceous age in an area in northern Colorado. The three stippled patterns represent sandstone members of the Pierre. The other mapped units are distinguished solely by the fossils the strata contain. The dashed lines indicate areas where fossil collections are sufficient to show a zone's presence but inadequate to locate its boundaries accurately. (After Scott and Cobban, 1965.)

The Pierre Shale in the area shown in Fig. 5-5 is not crowded with am-
monites. Most of them occur in calcareous concretions, some scattered, others
closely spaced in nearly continuous beds. The shale between concretion-bearing
beds contain few fossils, probably because of conditions unfavorable for their
preservation. Where concretion beds are locally covered by soil or alluvium the
boundaries of the range zones cannot be ascertained with confidence, and the
zones are represented on the map only by dashed lines known to fall within
their limits. Fossils other than those mapped also occur in the Pierre Shale, but
their distribution is not shown. Those whose range zones are shown have been
selected for abundance and inferred time-stratigraphic significance.

The species of *Baculites* mapped in Fig. 5-5 belong to three closely related
phyletic lineages. The earliest includes the species *B. obtusus* to *B. perplexus*,
the next includes the species *B. compressus* to *B. eliasi*, and the youngest includes
B. baculus to *B. clinolobatus*. The earliest representatives of each lineage segment
appear as migrants from other regions, but their partial ranges in the Pierre
are short and hence of great value. In each of the three lineages the succeeding
species evolves through transitional forms and its range zones are believed to
approximate the total ranges of these species in time.

Range zones of the three species of *Didymoceras* do not record a continuous
lineage, but only fragments of one. Range zones of these species, however, are
believed to approximate their total time ranges in this biogeographic province.
The upper two *Didymoceras* zones are separated by the range zone of *Exiteloceras jenneyi*, a migrant with no apparent ancestors or descendants in this region.
Its range zone is also believed to approximate its total time range in this province.

Species of *Baculites* are excellent for time-stratigraphic zonation. Their
lineage is well known. They evolved rapidly so that the total range in time of each
species is short. Moreover, they were swimmers and hence their distribution
was not controlled by environmental subtleties of the sea floor. Most of the
baculite species illustrated in Fig. 5-5 appear to be restricted largely to this
single biogeographic province. It contains a relatively complete record of their
evolution, which permits detailed correlation within the province. Because their
record is sparse outside this province, however, they are of limited value for
correlations with other provinces.

It should be re-emphasized that only a small minority of range zones
appear to represent the total time range of a species. Most range zones, at best,
represent a species' time range within a province. However, one must keep in
mind that such ranges are delineated empirically, and that the finding of the
species in a new area always carries the possibility of extending its known
stratigraphic range higher or lower.

Gradual Faunal Migrations

Normally a species migration into a given province requires only a geologic
instant once a barrier is overcome. However, the time significance of its initial

appearance must be inferred with caution. All barriers in geologic history were not breached suddenly. Apparently some were gradually swept out of the way.

A. R. Palmer has shown that the Late Cambrian migration of the Pterocephaliid trilobite fauna into the shallow shelf seas of the Cordilleran region (southern Nevada and adjacent areas) took a long time (see Fig. 5-6). From west to east the Pterocephaliid fauna appears in progressively younger strata. As the new fauna migrated eastward across the region, it slowly replaced the previously established Crepicephalid trilobite fauna. Much later the Pterocephaliid fauna itself was similarly replaced from west to east by the still younger Conaspid fauna. The basic stock of the Pterocephaliid and Conaspid lineages developed in another province, one that was separated from that of the Cordilleran region, not by a clear-cut physical barrier, but by one more subtle, possibly maintained by temperature. There, migrations of profoundly new faunas into a province produce time-transgressive faunal changes like those produced on a much smaller scale by the shifting of local environments.

This example does not invalidate the temporal value of most migrations. It simply reiterates that any *single* line of fossil evidence is subject to greater error than several lines interpreted collectively. For reliable interpretations all available evidence should be taken into account.

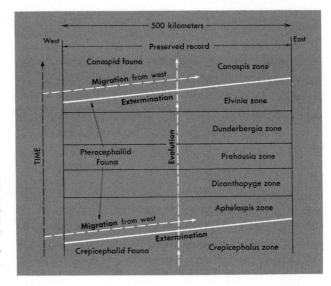

FIG. 5-6 The Pterocephaliid trilobite fauna slowly migrated eastward across the Cordilleran province in Late Cambrian time and replaced the established Crepicephalid fauna. Later the Pterocephaliid fauna was similarly replaced by the gradually migrating Conaspid fauna. (After A. R. Palmer, 1965.)

CONCURRENT RANGE ZONES

The narrowest time-stratigraphic zones with the most consistent boundaries provide the most accurate correlations. Logically, more precision results from utilizing the overlapping ranges of two or more species rather than the range zone of one species alone, because the probability of encountering an older or younger occurrence of several species at a new locality is much lower than that of encountering an older or younger occurrence of just one of them. Zones based on overlapping ranges have been called *concurrent range zones.* Zones of this kind were invented and extensively used (1856–58) by Albert Oppel in the Jurassic rocks of Europe, and no one since Oppel's time has been able to devise a more precise and reliable kind of time-stratigraphic fossil zone.

Oppel plotted the range zone of every species he found in numerous stratigraphic sections in a wide area. Some of the stratigraphic ranges were short, some were intermediate in length, and others were long. Species ranges tended to begin and end independently and somewhat randomly, rather than jointly, in each section. Not only did Oppel carefully note vertical ranges of each species, but where he could he also inferred phyletic lineages. He placed the boundaries of his zones at the first appearances of a new lineage. Other selected species were equally important, however, for definition of the zone; some first appeared within it, others terminated within it. The fossils selected to depict a concurrent range zone were thus chosen for maximum reliability in time-stratigraphy. Typically many long-ranging species also occur in a given zone, but since they also occur in zones below and above, they are not included in the definitive association.

Concurrent range zones may include fossils from deposits formed in many environments, and in such cases their boundaries tend to be independent of minor changes in depositional environments. By the same token some of the characteristic species may never actually occur together in the same bed. A concurrent range zone is named after one of its species, but that species has no more significance in recognizing the zone than does any other single member of the diagnostic association. The name-giver need not be limited to the zone and it need not occur throughout the zone. Oppel pointedly noted that zones could just as well have been named after localities.

Most concurrent range zones can be recognized only within a single biogeographic province. The minor events of migration and evolution on which they are based typically have no synchronous counterparts in other provinces. When characteristic zonal species that have evolved within a province happen to spread suddenly into another province, they can provide invaluable tie points between the independent zonal sequences of each, but exact zonal *boundaries* can rarely be carried from one province to another.

Although most fossil zones are too narrowly defined to be recognizable beyond their provinces, correlation can be established among provinces by utilizing groups of zones, or stages. Stages are recognized in the same way as zones; the difference is in refinement. Stages are based on larger faunal changes than zones and they can be recognized over the geographic extent of adjacent provinces; some can even be recognized worldwide. On a global basis, however, stages generally present the same difficulties as do zones when traced beyond their provinces. Most are simply refined too highly for worldwide extension. Hence, they are lumped into still larger units—series—which are recognized on even more gross organic changes than stages. With local exceptions, series can be traced fairly confidently all over the world.

Methods of Selecting Zone Boundaries

Workers since the time of Oppel have adopted various means of defining concurrent range zones to suit their special needs. Some paleontologists have selected boundaries that encompass certain combinations of bottoms and tops of ranges. Others, impressed with the very large numbers of particular species in some beds and their relative scarcity in others, have attempted to weave the idea of peak abundance into the concept of concurrent ranges. Still other paleontologists, dealing with subsurface information obtained by drilling, have focused their attention particularly on the upper limits of range zones. They do so partly because the geologic record is unfurled in reverse order in drilling, and partly because down-hole cavings obscure the lower limits.

Figure 5-7 illustrates four ways in which fossil ranges in one section could be divided into concurrent range zones. The scheme that Oppel might have selected, and which most European workers prefer today, is represented by

FIG. 5-7 Four methods of recognizing concurrent range zones in a single section. Width of lines indicates the relative abundance of specimens. (Modified from Schenck and Graham, 1960.)

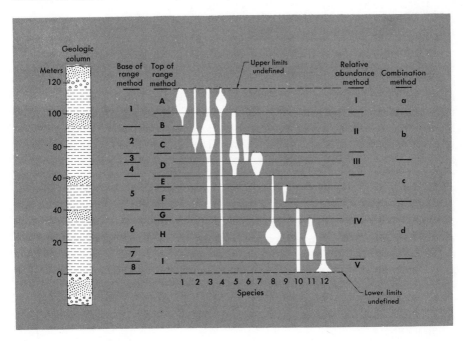

Arabic numbers on the left side of the range chart. In this method a zone is bounded below by the lowest appearance of a significant species (preferably representing a new lineage), and above by the lowest appearance of another significant new species. In addition, each zone contains within it the first appearance of some species and the last appearance of others, and these aid in its recognition.

A method that might be applied by a subsurface geologist is shown by the capital letters on the left side of Fig. 5-7. One that takes abundances of each species into account is shown by the Roman numerals on the right side. Finally, a method designed to include as many species as practicable within each zone is shown by the lower case letters in the far right column of Fig. 5-7. The last method includes more lower and upper limits of ranges within each zone than the others do, and extensions of ranges in new areas tend to disrupt these zonal boundaries less than with the other three methods.

None of these methods is consistently superior to the others. In theory Oppel's approach, which emphasizes lower limits of range zones, may be the most sound, but in subsurface work where bases of zones may be impossible to ascertain within acceptable limits of accuracy, the use of upper boundaries may be the only method that works. In very small areas, zones based on abundances may actually be the most accurate. The method of zoning should depend on the information that is necessary to the task at hand and that can be collected under the circumstances.

INDEX FOSSILS AND PEAK ZONES

The term "index fossil" suggests to most geologists a fairly abundant, distinctive species, narrowly limited in time and widely distributed geographically. Species that actually possess these attributes are valuable aids in correlation. So-called index fossils typically appear in the stratigraphic record fully formed with no indication of ancestry. This is one reason they are distinctive, but lineages of such species are poorly known. The ranges that are actually utilized are simply range zones, which, of course, may represent different durations of time in different areas. In correlation based solely on such index species there is no independent way of checking whether or not the stratigraphic range at a distant locality is the same as that known previously. Although many so-called index fossils appear to be reliable time indicators, others have been found to be unreliable. This lack of "quality control" has narrowed the use of index fossils considerably in modern paleontologic correlations.

Index fossils were initially conceived as a shortcut in correlation. The impetus for seizing upon particular index fossils may have actually stemmed from Oppel's method of naming concurrent range zones after just one of its representative species. Although Oppel had emphasized that each of the several species in the concurrent range zone association was equally significant, the one whose name the zone bore became somehow special to early geologists. The single species idea took root and by the end of the nineteenth century, paleontologists had acquired the habit of designating a zone's most abundant and characteristic species as "index." This species was supposed to represent the zone, and it came to be considered, particularly among nonpaleontologists, the key to the zone's recognition. The common practice of relying on just one species generally promoted reliance on simple range zones rather than concurrent range zones.

Later the term "index fossil" took on another undesirable connotation, particularly in the United States where it came to be thought of as a species that was characteristic of a particular formation. Saying that "this species is index for that formation" implies that a fossil found anywhere in the formation will be present throughout. This practice also implies that, because the index fossil has time value, the formation also has time value, thus confusing rock units and time-stratigraphic units and lending credence to the old fallacy that formations are time-parallel.

If a fossil species ranges throughout a formation in a given region, the species provides no events indicative of time. Probably it merely accompanied the environment that produced the formation's distinctive lithology. The species might aid mapping if the formation is one of several in the area that have similar lithologies, but this is not time correlation; it is only a means of identifying a part of the local section. To establish time correlation we must seek species that do not range throughout a formation, but which appear or disappear as a result of time-significant evolutionary or migration events. Thus, it is most important that stratigraphic occurrences of fossils be reported accurately *within* formations.

Another concept, closely allied to the concept of index fossils, was the idea that a great abundance of a fossil species could be interpreted as having time significance. In this view, it was not the organism's range, but its time of flowering, or "peak abundance," that provided the basis for a time-stratigraphic unit. Now, if there were some point in the history of a species when numerical abundance was a distinct maximum, and if this abundance were actually recorded everywhere in the strata that contain it, then the use of "peak zones" as time-stratigraphic units would be theoretically sound. In reality, however, maximum abundance may be reflected in some areas and not in others. The strata containing the most specimens may represent one of several things: sporadically favorable environments, sudden unfavorable conditions that resulted in mass mortality, a time of exceptionally slow sedimentation but sustained organic productivity, or mechanical accumulations of dead shells.

QUANTITATIVE BIOSTRATIGRAPHIC METHODS

Graphing the Correlation Line

Alan Shaw has proposed a correlation technique that utilizes the appearances and disappearances of all the species which two stratigraphic sections have in common. First, the range of zones are carefully established in the measured sections. Then the bases and tops of the range zones are plotted on a graph (see Fig. 5-8), using section A as the horizontal axis and section B as the vertical axis. These points tend to cluster around a "correlation line," like that shown in Fig. 5-9, whose intercepts depict precise correlation between the two sections. This kind of diagram is thus a good stratigraphic tool because it takes into account every potentially time-significant event in the two sections. In applying it one need have no concern for the often arbitrary selection of zone boundaries.

The points in Fig. 5-9 that fall well off the correlation line represent species which appear or disappear in one section much earlier than in the other. Such species are strongly facies-controlled. Hence judgments as to which fossils will be useful for biostratigraphic correlation and which ones will not can be made easily and objectively once the diagram is plotted. Physical events can also provide points for the correlation line. A bentonite bed that occurs in both sections represents an identical time level. This helps greatly in locating the correlation line, which should pass directly through it (see Fig. 5-9).

FIG. 5-8 (Left) Schematic representation of the sections shown in Fig. 5-9. (Right) Another section with an identical time span and identical fossil content but only half the rate of rock accumulation. Small numbers at limits of range zones indicate stratigraphic positions in meters above base.

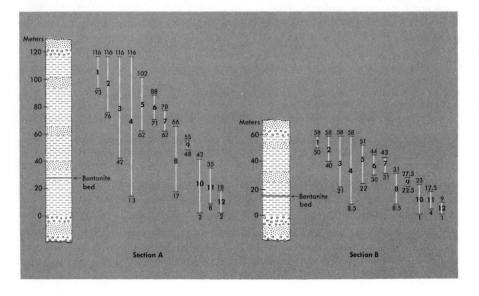

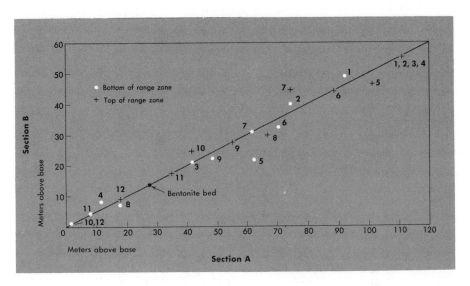

FIG. 5-9 Correlation line of the sections shown in Fig. 5-8.

Changes in slope of the correlation line tell much about the geologic history of an area. If the rate of sedimentation at one section increased or decreased significantly, the result is an expanded or condensed interval, and this shows up as a doglegged correlation line (see Fig. 5-10). A hiatus appears as a horizontal or vertical segment in the line (see Fig. 5-11).

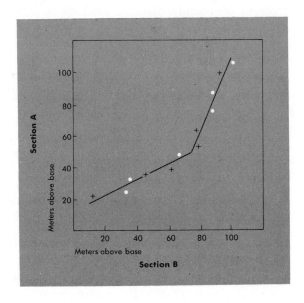

FIG. 5-10 Dog-legged correlation showing change in sedimentation rate.

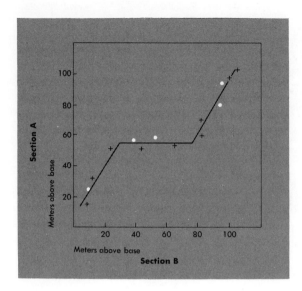

FIG. 5-11 A hiatus in Section Y appears as a horizontal line which offsets the correlation line.

Coiling Ratios in Planktonic Foraminifera

Planktonic foraminfera make their many-chambered shells out of calcite. Most are trochospirally coiled; that is, they coil in a spiral with all the chambers visible from one side of the shell and only those of the last whorl visible from the other side (see Fig. 5-12). Individuals belonging to a given species may coil either to the left or to the right. Some existing species of planktonic foraminifera occur in dominantly right- or left-handed populations, depending upon the temperature. For example, *Globorotalia truncatulinoides*, which is widespread in today's oceans, is dominantly right-handed in areas of warm waters and left-handed in areas of cold water. At times during the Pleistocene Epoch, when glaciers periodically covered much of the land areas of the Earth, the oceans cooled considerably. During these times dominantly right-coiled populations of *G. truncatulinoides* were replaced in middle and low latitudes by dominantly left-coiled populations.

These coiling changes provide a correlation tool capable of resolving Pleistocene climatic events that occurred over time periods of tens of thousands or hundreds of thousands of years, an order of magnitude less than the time that can be resolved using conventional zoning criteria. Figure 5-12 shows *G. truncatulinoides* coiling ratios in three cores that recovered Atlantic Ocean bottom sediments deposited over the last 1.5 million years. The proportion of right- and left-handed specimens fluctuates similarly throughout all three cores, and provides the basis for correlations. This method of biostratigraphic correlation is applicable only to a part of a given ocean basin, and even in a small region it is best supplemented by other correlation tools. Nevertheless, it provides a precision for the Pleistocene that surpasses that of most faunal zones.

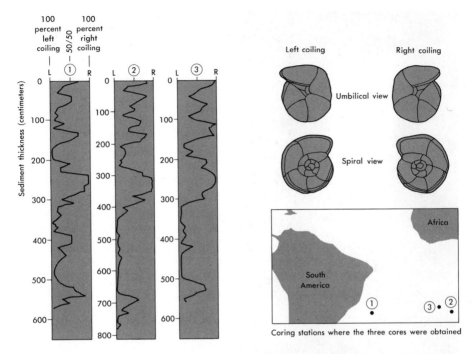

FIG. 5-12 Coiling ratios of *Globotruncana truncatulinoides* in three South Atlantic Ocean cores. Correlations are apparent for the 1.5 million years represented. (Data from Ericson and Wollin, 1968.)

A Quantitative Paleontologic Indication of Geologic Time

This final example does not deal with a correlation tool, but with a rare quantification of geologic time using fossil shells. It is based on the gradual decrease in the number of days in the year throughout geologic time.

Astronomers have long inferred that the Earth's rate of rotation must be slowly decreasing due to friction of the tides. Tidal energy comes from the rotation of the Earth itself, and because this energy is spent largely in the work of ocean currents, which must finally produce heat, the overall effect is like the everyday drag of friction that slows a rotating wheel. Although tidal energy is huge by civilizations' standards, it is extremely small compared with the kinetic energy of the Earth, so small that its retarding effects have only recently been measured. Tidal retardation is causing the length of day to increase by about two seconds every 100,000 years. At this rate the length of day at the beginning of the Paleozoic Era—570 million years ago—would have been about 21 hours. Meanwhile, the period of revolution about the sun would be expected to remain constant. A year was just as long as it now, but because the day was shorter it contained more days, about 420 at the beginning of the Paleozoic.

FIG. 5-13 A specimen of the Silurian coral, *Holophragma calceoloides,* with a total of 240 striations interpreted to represent 240 days' growth. (Magnified four times.) (Courtesy J. W. Wells.)

John Wells, an American paleontologist, has found evidence for both daily and yearly growth increments in Paleozoic corals. Wells suggested, on the basis of modern counterparts, that the prominent ridges girdling many Paleozoic horn corals represent yearly rings. Similar ridges are known to be produced yearly in modern corals and this inference appears to be sound. Thus, the age of a horn coral specimen can be ascertained by counting the ridges that encircle it. In addition to these ridges, well-preserved horn corals also show extremely fine, closely spaced striations that encircle the coral on the outer surface (Fig. 5-13). These striations, too, almost unquestionably represent growth increments. In modern corals similar increments occur daily, and the same was probably true for fossil corals. Therefore, by counting the number of fine ridges between the large yearly growth rings on the Paleozoic corals, Wells was able to infer the number of days in a year at the time the coral lived.

Only exceptionally well-preserved fossil corals show the fine striations, and even in the same specimen growth rates vary with short-term environmental fluctuations, thus scattering growth-line counts. But in three coral genera from Middle Devonian rocks of New York and Ontario, the numerous counts range between 385 and 410 and they cluster near the median of 398. The scatter is shown by the vertical bar A in Fig. 5-14, and the indicated age of about 380 million years is essentially the same as that obtained by radiometric dating of rocks of this age. Corals from the Pennsylvanian of two different regions of the United States indicate 390 and 385 days per year (bar B, Fig. 5-14), only slightly fewer than the 392 days expected for rocks formed 300 million years ago, the radiometric age determined for the mid-Pennsylvanian.

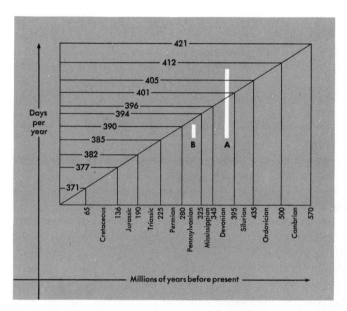

FIG. 5-14 Predicted number of days per year plotted against quantitative time scale. Vertical bars show results from daily striations and annual rings on fossil corals. (From J. W. Wells, 1963.)

In summary, the ages in years indicated by the fossil corals agree remarkably well with radiometric ages in years obtained from rocks of the same age. This verifies that radiometric dates, which are our conventional source of quantitative ages, are in the right ball park. This is particularly reassuring because the verification comes from a biologic system—a source of support as different from that of radioactive minerals as could be envisioned. These results suggest that the quantitative ages yielded by radiometric dating are reliable in general terms. Numerous cross-checks using various radiometric dating methods are internally consistent and further attest to the accuracy of the radiometric dating process. The following discussion in Chapter 6 will briefly explain these radiometric methods.

six

radiometric dating

Many kinds of atoms that occur in nature are unstable and they change spontaneously to a lower energy state by radioactive emission. This process is termed *radioactive decay*. During decay one kind of atom called the *parent* changes into another kind of atom called the *daughter*. A given kind of atom is distinguished from all others by the number of protons and neutrons in its nucleus. The number of protons determines the *element* to which the atom belongs. The number of protons and neutrons together give the atom its mass. Thus the proton number and mass number together specify a single, particular kind of atom. For "particular kind of atom," we may use the convenient word *nuclide*. Two different nuclides that contain the same number of protons but different numbers of neutrons belong to the same element but have different masses. These are referred to as *isotopes* of the same element. For example, two common isotopes of uranium (proton number 92) are uranium-235 and uranium-238.

In the process of radioactive decay the nucleus of a parent radioactive atom emits an alpha particle or a beta particle, or it captures an electron. In alpha decay the nucleus of the parent atom loses 2 protons and 2 neutrons; the mass number thus decrease by 4 and the proton number decreases by 2. In beta decay, the nucleus emits a high-speed electron, and one of its neutrons turns into a proton; the proton number increases by 1 but mass remains unchanged. In electron capture, a proton in the nucleus picks up an orbital electron and turns into a neutron, thus decreasing the proton number by 1. Again mass remains unchanged.

HOW RADIOMETRIC DATING WORKS

Each radioactive nuclide has one particular mode of decay and, more importantly, its own unique rate of decay, which is a constant unchanging property of the nuclide. Radioactive decay takes place entirely in the atomic nucleus and the rate of decay is independent of external conditions, such as heat or pressure. The rate is even unaltered by chemical changes such as oxidation or reduction of the parent atom because these involve only the orbital electrons and not the nucleus. Hence, if a radioactive nuclide is incorporated into a mineral or rock when it crystallizes, the amount that decays to the radiogenic daughter nuclide is controlled only by the elapsed time since the crystallization event.

The principle of radiometric dating is comparable to an hourglass. Turn the glass over and sand runs from the top chamber to the bottom. So long as some sand remains in the top, the amount in the top relative to the amount accumulated in the bottom provides a measure of the time that has elapsed. The sand in the top of the hourglass represents decaying radioactive atoms and that in the bottom accumulating daughter atoms. Just as the hourglass must be sealed so that sand can't escape through the sides, so the atomic lattice structure of the mineral must be able to hold both the parent and the daughter atoms without allowing any to escape, or, for that matter, to enter from an external source. In other words, the system must be closed.

Unlike the passage of sand through an hourglass the uniform shrinking of a candle, or other *linear* rates of depletion, radioactive decay occurs at a *geometric* rate (see Fig. 6-1). Each individual atom of a given radioactive isotope has the same probability of decaying within the next year, and this probability remains the same no matter how long the material being dated has been in existence. Probability of decay is expressed by a number called the *decay constant* λ which simply stipulates the *proportion* of atoms of that particular nuclide that always decay in a year. The actual *number* of atoms that will decay is λN, where N is the number of radioactive parent atoms present in the system at the beginning of the year. At the beginning of the next year, the number of radioactive parent atoms is, of course, smaller, having decreased by λN. Hence the actual number of atoms to decay the second year is smaller, and the number decreases with each successive year. The total time required for *all* the radioactive atoms in a given system to decay cannot be specified. In theory it is infinite. It is a simple matter, however, to specify the time required for half of the atoms of a particular radioactive nuclide to decay. This time period is called the *half-life*. Each radioactive nuclide has its own half-life; some have a duration of microseconds, others, trillions of years. A given nuclide is best suited to measure ages that are about the same order of magnitude as its half-life.

The end of one half-life period marks the beginning of another. Thus if a quantity of a radioactive nuclide is segregated, half of the initial number of atoms

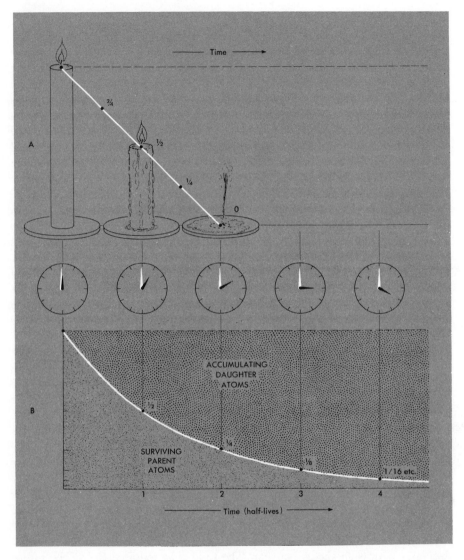

FIG. 6-1 (A) Uniform straight line depletion of most everyday processes. (B) By contrast, the radioactive decay curve approaches zero line asymptotically. The end of one half-life interval is the beginning of a new one.

$(N_0/2)$ remains after one half-life period, half of those or one-fourth $(N_0/4)$ remains at the end of the next half-life period, and half of those or just one-eighth $(N_0/8)$ remains at the end of the next half-life period, and so on. The number of surviving parent atoms (N_t) at the end of the many (n) half-lives is simply $N/2^n$. Plotted as a function of time t the surviving parent atoms express a curve like that shown in Fig. 6-1(B). This simple relationship is the basis of all radioactive clocks. The equation for this kind of curve is $N_{(t)} = N_o e^{-\lambda t}$, in which e is 2.718.

When mineral grains containing radioactive atoms first crystallize as part of a newly formed rock, they ideally contain no atoms of the radiogenic daughter. The initial daughter/parent ratio is zero and therefore the indicated age is zero. With time, the decay of radioactive parent atoms produces radiogenic daughter atoms in their places in the mineral grains. Knowing the decay constant of the radioactive parent, we only need measure the ratio of radiogenic daughter and parent nuclides (D/P) in the mineral to calculate the time, measured in years before present, that the system originated. This time is called the *radiometric age* of the system. It is calculated from the equation $t = (1/\lambda)\log_e[(D/P) + 1]$. Figure 6-1 shows graphically the logic of radiometric dating at a glance. Clearly, with each increment of time, the parent/daughter ratio is unique.

Error in Radiometric Dating

Two assumptions are implicit in the radiometric dating of a mineral or rock; first, that the system has remained closed—that is, neither parent nor daughter atoms were added or removed other than by radioactive decay; second, that no atoms of the daughter nuclide were present in the system when it formed. Many minerals and rocks satisfy the closed system requirement well enough, but fewer were originally entirely free of daughter nuclide. In many cases some atoms of daughter nuclide were accidentally incorporated as contaminants when the mineral to be dated crystallized. This is called *original daughter*, and its presence must be detected. If it is lumped with the *radiogenic daughter*, which has formed as a product of radioactive decay through time, then the calculated age of the mineral will be greater than the actual age. Fortunately there are ways of estimating original daughter, and hence it is not generally a large problem in age determination.

Decay constants for nuclides widely used in radiometric dating are known within one or two percent. Besides possible small *systematic* errors due to these uncertainties, there are two chief sources for error in any radiometric age: (1) partial loss of radiogenic daughter and (2) errors in the laboratory analysis. Examples of both will be discussed in the sections to follow. The point to be made here is exactly what the stipulated error means in a radiometric date. For example, in a date such as 325 ± 10 million years, the "plus or minus ten" means only that the age *determination* is probably reproducible within this 20-million-year range. This range refers to the precision of the measurement, that is, its repeatability. Usually the error given is the standard deviation of the scatter of dates obtained in making several analyses. This scatter of dates is due to (1) error in laboratory measurements and (2) variations among samples from the same system. The meaning of 325 ± 10 million years, then, is simply this: should another sample of the same rock be analyzed, the chances are good that it will yield a date within this 20-million-year time range.

Accuracy, as opposed to precision (to use statistical jargon), refers to the deviation of the determined age of the rock from its true age. The greatest source of inaccuracy in geochronometry is the failure of rocks and minerals to remain closed systems. Loss of the daughter product can be detected only by checking results of more than one dating method. Thus the best evidence for good accuracy is concordant results from different methods.

Radioactive Nuclides in Nature

It is convenient to recognize two categories of radioactive nuclides, long-lived nuclides with half-lives of several hundred million years or more, and short-lived nuclides with half-lives of less than a few tens of millions of years. Of the multitude of radioactive nuclides that were present in the Earth when it formed, only the 21 long-lived nuclides have survived from that primordial time. The many short-lived radioactive nuclides that were part of the original Earth are extinct, having decayed to undetectable quantities long ago. However, several short-lived radioactive nuclides do exist naturally on Earth today because they are continually produced either as one of the steps in the uranium or thorium decay chains, or by cosmic ray bombardment, chiefly in the upper atmosphere. The approximately 10,000 tons of cosmic dust that the Earth

Table 6-1 The Chief Methods of Radiometric Age Determination

Parent Nuclide	Half-life (years)	Daughter Nuclide	Minerals and Rocks Commonly Dated
URANIUM-238	4,510 million	Lead-206	Zircon
			Uraninite
			Pitchblende
URANIUM-235	713 million	Lead-207	Zircon
			Uraninite
			Pitchblende
POTASSIUM-40	1,300 million	Argon-40	Muscovite
			Biotite
			Hornblende
			Glauconite
			Sanidine
			Whole volcanic rock
RUBIDIUM-87	47,000 million	Strontium-87	Muscovite
			Biotite
			Lepidolite
			Microcline
			Glauconite
			Whole metamorphic rock

collects each day brings additional short-lived nuclides, also produced by cosmic ray activity. These short-lived radioactive nuclides are present in vanishingly small amounts, but a few are useful thanks to highly refined analytical techniques. By far the most widely used is carbon-14, which is a valuable dating tool for the last 40,000 years of Earth history. Other short-lived nuclides have found special use in dating deep-sea sediments as old as a few hundred thousand years.

Most of the *long-lived* radioactive nuclides that occur in nature are too rare to be of wide use in dating rocks, and others have half-lives that are too long for the dating of even the oldest rocks. Just four nuclides, potassium-40 rubidium-87, uranium-235, and uranium-238, have provided almost all of the radiometric ages for ancient rocks. The age determination methods based on these nuclides are summarized in Table 6-1. Thorium 232 has also provided some important ages, but since it typically occurs with uranium it has been mostly of corroborative value.

POTASSIUM-ARGON METHOD

Potassium is the seventh most abundant element in the Earth's crust, and 0.4 percent of all potassium is radioactive potassium-40. Because potassium is a constituent of common rock-forming minerals, such as the micas, potash feldspar, and hornblende, the potassium-argon method has proven valuable in dating many kinds of rocks. Only 11 percent of potassium-40 decays to argon-40 (by electron capture); the rest decays to calcium-40 (by beta decay). The decay of potassium to calcium does not provide a workable dating tool because some calcium occurs in virtually all common rocks and minerals, and the radiogenic calcium cannot be discriminated from original calcium. Argon, however, being an inert gas, is never bound chemically in newly formed potassium minerals. Any argon-40 in the above minerals is almost certainly a result of decay in place from potassium-40. In rare instances igneous rocks have trapped original argon mechanically but this is not a major problem. Hence the technique of potassium-argon dating has become an excellent chronologic tool.

Igneous extrusives and other rocks that have never been buried deeply commonly give reliable potassium-argon ages. However, above temperatures normally attained at a depth of around 5 kilometers some of the argon gas escapes from crystals in which it is produced. Igneous and metamorphic rocks do not begin to retain all of their argon until they have cooled to 200°C or so. If they crystallize at great depth this temperature may not be reached until several million years after crystallization. For this reason potassium-argon ages are usually considered to be minimum ages.

Figure 6-2 indicates how extensive the argon loss due to heating can be and how it differs from one mineral to the next. About 55 million years ago Eldora Stock intruded the 1,300 million-year-old Precambrian Idaho Springs

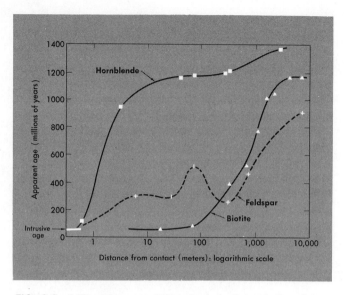

FIG. 6-2 A 55-million-year-old intrusive in the Colorado Front Range caused argon loss and resulting mixed potassium-argon ages in the minerals of the adjacent 1,300-million-year-old metamorphic rocks. (After S. R. Hart, 1964.)

Schist in the Colorado Front Range. Minerals of the Idaho Springs within a few centimeters of the stock did not melt, but they lost all of their argon and now record the stock's age. Minerals of the Idaho Springs several kilometers away retained all of their argon and they record their original Precambrian age. At intermediate distances the argon loss was partial, resulting in mixed ages, which fall between the age of the Idaho Springs and the age of Eldora Stock. Mixed ages such as these are misleading because they record the age of neither event. Figure 6-2 shows that hornblende lost argon the least, biotite and feldspar the most. Thus even if one were unaware of the regional intrusives, a problem would be indicated by the lack of concordance among the various minerals analyzed.

Potassium-Argon Whole Rock Dating

Dating based on a specific mineral like biotite or hornblende requires the picking of individual mineral grains from crushed rock to obtain enough for analysis. Many rocks are so fine-grained that individual grains can be separated only with great difficulty. These rocks are commonly analyzed without separating out individual mineral components. This procedure is called the *whole rock method*. With a half-life of 1,300 million years, potassium-argon dating is applicable to the oldest Precambrian rocks we know. But tiny amounts of argon can be measured with great precision and the potassium-argon whole rock method has been applied successfully to volcanic rocks as young as 100,000 years. Whole rock dating of volcanic basalts of the Hawaiian Islands shows that the main islands grew progressively from Kauai on the northwest, where the last volcanism occurred 3.8 million years ago, to Hawaii on the southeast, where two volcanoes, Mauna Loa and Kilauea, are still active (see Fig. 6-3). This result quantifies a trend long recognized from geomorphic evidence alone.

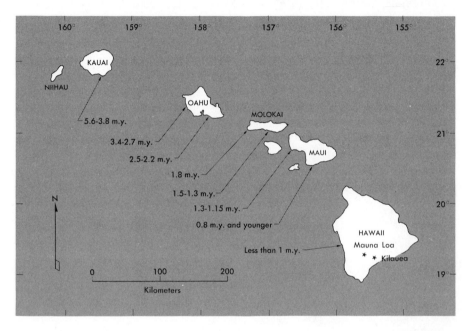

FIG. 6-3 Potassium-argon ages of the major Hawaiian Islands. The volcanic activity which formed the islands migrated progressively from northwest to southeast. (After I. McDougall, 1964.)

RUBIDIUM-STRONTIUM METHOD

Radioactive rubidium-87 decays in a single beta step to strontium-87. Rubidium-87 constitutes 28 percent of all rubidium, but rubidium is not very common; the Earth's crust contains only about 15 percent as much rubidium-87 as potassium-40. Rubidium occurs in trace amounts in micas and potassium feldspars, pyroxenes, amphiboles, and olivine, all of which may be used for rubidium-strontium age determinations. A trace amount of original strontium almost always occurs in these minerals, and it must be estimated routinely as part of any rubidium-strontium age determination. The quantity of original strontium-87 is then subtracted before calculating the age.

Original strontium always contains nonradiogenic strontium-86, whose quantity remains constant. The proportion of strontium-87 to the nonradiogenic strontium-86 was identical throughout the rock body when it formed. Hence the original strontium-87/strontium-86 ratio can be used to calculate the quantity of original strontium-87 in the radioactive sample. To determine the strontium-87/strontium-86 ratio that prevailed when the rock formed, one first analyzes a rubidium-*free* sample. Next a rubidium-*rich* sample is analyzed, not only for rubidium-87 and strontium-87, but also for strontium-86. Using the original strontium-87/strontium-86 ratio one calculates the quantity of strontium-87 that is original and the quantity that is radiogenic (see Fig. 6-4). The age of the sample is then determined from the daughter-parent ratio in the usual way.

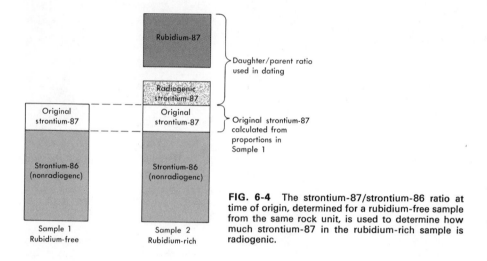

FIG. 6-4 The strontium-87/strontium-86 ratio at time of origin, determined for a rubidium-free sample from the same rock unit, is used to determine how much strontium-87 in the rubidium-rich sample is radiogenic.

This method works best for rocks that are geologically old. The extra analytical step to determine the original strontium-87/strontium-86 ratio introduces extra analytical error, which may be huge if the sample is young. For example, suppose a biotite sample contains 14 parts per million (ppm) original strontium, which includes one ppm strontium-87, the original daughter. The biotite also contains 250 ppm rubidium-87, the parent radioactive nuclide (these are typical values for biotite). The rubidium-free sample is analyzed for the original strontium-87/strontium-86 ratio (see Fig. 6–4) with an error of one percent. This ratio is then used to calculate the original strontium-87 in our biotite sample, where a one percent error applied to one ppm strontium-87 gives an accuracy of one part in 100 million, which is not bad. However, one part in 100 million is *all* the radiogenic strontium that the rubidium-87 in our biotite will produce in 3 million years. Hence the potential error for such a young sample, from this analytical step alone, is 100 percent (see Fig. 6-5). However, if the age of the sample were 1,000 million years (late Precambrian) the rubidium-87 in it would have generated 3.7 ppm radiogenic strontium-87.

FIG. 6-5 Dating error de-
creases with sample age. Chart
assumes a one percent error in
measuring original strontium-87
for biotite containing one ppm
original strontium-87 and 250
ppm rubidium-87.

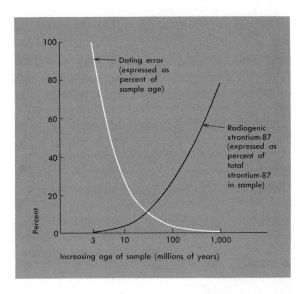

FIG. 6-5 Dating error de-creases with sample age. Chart assumes a one percent error in measuring original strontium-87 for biotite containing one ppm original strontium-87 and 250 ppm rubidium-87.

In this case the analytical error becomes an insignificant 0.3 percent (Fig. 6-5). This is why rubidium-strontium dating, and uranium-lead dating as well, are excellent for ancient rocks but not very reliable for young rocks.

A widely applicable means of minimizing the analytical error inherent in rubidium-strontium dating is the *isochron method*. Several samples are required but none need be rubidium-free or unusually rubidium-rich. The isochron method requires only that the samples vary in relative content of rubidium. At the time of crystallization the strontium-87/strontium-86 ratio was the same in all parts of the rock body and a group of samples with varying rubidium content would fall on a horizontal line on a graph that has the strontium-87/strontium-86 ratio the ordinate and the rubidium-87/strontium-86 ratio the abcissa (see Fig. 6-6). With time the rubidium-87 content decreases and the strontium-87 content correspondingly increases in each sample. On the graph each sample thus follows its own path upward and to the left, each moving at a rate proportional to its relative rubidium content. The series of samples continues to define a straight line whose slope is a function of elapsed time since crystallization. This line is the *isochron* (same time) and its slope provides the radiometric age of the rock. (Its intercept on the ordinate additionally provides the original strontium-87/strontium-86 ratio.)

The rubidium-strontium method is useful for both igneous and meta-morphic rocks. It has successfully dated Moon rocks collected by the Apollo missions. For example, Fig. 6-6 shows an accurate rubidium-strontium isochron age of 3,300 ± 80 million years for lunar basalt from the edge of Hadley Rille. The isochron is defined by separate minerals from the basalt as well as by whole

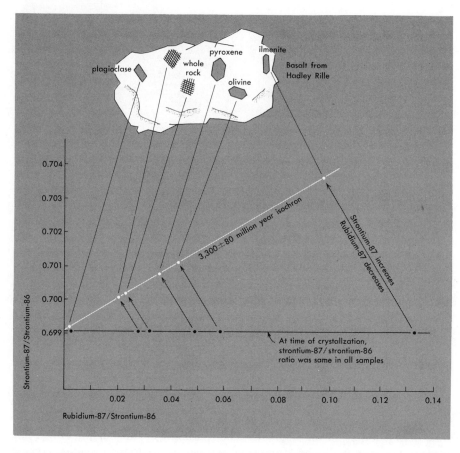

FIG. 6-6 Rubidium-strontium isochron dating of a lunar basalt. The isochron method depends on analysis of samples with differing rubidium content. Slope of the isochron increases with time and provides the date. (Data from Murthy and others, *Science,* v. 175, 1972, p. 419–421.)

rock determination of mixed mineral aggregates. This age is verified by potassium-argon dating. Volcanic rocks of this age have been found at other lunar localities and they may represent an important igneous episode on the Moon.

On Earth the rubidium-strontium method is perhaps the most valuable technique available for dating metamorphic rocks. Strontium is easily mobilized during metamorphism and the strontium-87/strontium-86 ratio thereby homogenized in the rock body, satisfying the requirements of the rubidium-strontium isochron method. Then as the rock body cools, the strontium becomes stabilized in minerals quickly and at much higher temperatures than does argon gas. Thus for metamorphic rocks the method generally yields better results than the potassium-argon method.

URANIUM-LEAD METHOD

All naturally occurring uranium contains radioactive uranium-238 and radioactive uranium-235 in a ratio of 138:1. Uranium-238 decays to lead-206 and uranium-235 to lead-207; hence these two separate nuclides provide a cross-check in determining ages. Most uranium minerals also contain radioactive thorium-232, which decays to lead-208, and this method is occasionally utilized as still another cross-check. The uranium and thorium nuclides decay through a *series* of intermediate nuclides that are themselves radioactive before finally arriving at their stable lead daughters. The uranium-238/lead-206 series involves eight alpha decay steps and six beta steps; the uranium 235/lead-207 series involves seven alpha steps and four beta steps; the thorium-232/ lead-208 method involves six alpha steps and four beta steps. Helium, which is produced in the alpha decay steps of each series, is an additional stable by-product.

Minerals that contain uranium as a chief component are rare, but minerals that contain uranium in trace quantities are fairly common. The most useful of these is the igneous mineral zircon ($ZrSiO_4$), which typically contains about 0.1 percent uranium. Small quantities of zircon occur in granitic rocks of many ages, and thus uranium-lead dating is widely applicable. Uranium-lead dates have also been achieved by whole-rock analyses of special rock suites, including samples from the Moon.

Original lead in a uranium-bearing mineral causes the radiometric age to exceed the true age unless it is detected. Lead-204, which is never produced radiogenically, provides a convenient way of detecting it. Lead-204 constitutes a small proportion of all common lead. If it is present, then other isotopes of lead, including lead-206 and lead-207, were present when the mineral formed. Knowing the isotopic composition of common lead at the time the minerals formed, the lead-204 is used to calculate the quantities of original lead-206 and lead-207 so that these can be subtracted in calculating the radiometric age.

After making allowance for original lead, the uranium-235/lead-207 and uranium-238/lead-206 ages should agree, provided the mineral has remained a closed system. If they do agree, the ages are said to be *concordant* and the probability is high that the radiometric age is the same as the true age. On a graph with the uranium-235/lead-207 ratio the ordinate and the uranium-238/lead-206 ratio the abcissa, the loci of all concordant ages define a curve called the *concordia*. If the uranium-lead ages do not agree then they do not fall on the concordia and they are said to be *discordant*.

An example of concordant ages is shown in Fig. 6-7. Uranium-lead dating for breccia and fine surficial material collected from the Moon by Apollo 11 gives ages between 4,600 and 4,700 million years. A corroborative age of 4,650 million years is obtained by thorium-232/lead-208 dating. This lunar material appears to represent a closed system, which originated between 4,600

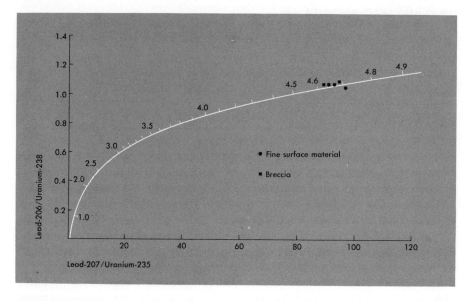

FIG. 6-7 Uranium-lead concordia showing the essentially concordant ages of lunar fine surficial material and breccia collected by Apollo 11. Age between 4,600 and 4,700 million years accords well with age of solar system. (Data from Wetherill, 1971.)

and 4,700 million years ago and which has not been significantly disturbed since. The 4,600 to 4,700 million-year date agrees with ages determined from meteorites. This is inferred to be the age of the Moon and of the rest of the solar system as well.

Discordant ages are caused by loss of lead from uranium-bearing minerals Daughter lead atoms are chemical misfits in crystals in which they have been produced, and hence they tend to escape selectively when the crystal is subjected to heat or stress. This always results in uranium-lead ages that are younger than the actual age, and as discordant ages that plot below the concordia. If different quantities of lead were lost from different parts of a rock body during a *single episode* of heat or stress, then samples from these different parts will plot, not as a single point, but as a straight line below the concordia. The line thus defined will intercept the concordia at two points and these two points represent (1) time of initial crystallization and (2) time of later lead loss.

An example is shown in Fig. 6-8. Seven discordant zircon analyses from late Precambrian volcanic and granitic rocks from the Blue Ridge of the Appalachian Mountains define a line with concordia intercepts of 240 and 820 million years. This documents an episode of intrusion and volcanic activity some 820 million years ago in the late Precambrian. The lead loss at 240 million years occurred during the orogeny that created the Appalachian Mountains in the late Paleozoic. Discordant ages, in this case, provide more information than would a single suite of concordant ages.

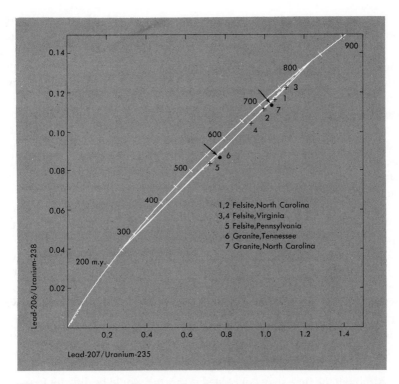

FIG. 6-8 A concordia plot of a series of discordant uranium-lead ages from the Appalachian Mountains indicates a major igneous episode 820 million years ago and the subsequent Appalachian Orogeny 240 million years ago. The curve is an expanded portion of a small segment of the curve in Fig. 6-7. (Data from Rankin and others, 1969.)

METHODS FOR THE RECENT GEOLOGIC PAST

Carbon-14 Method

Carbon-14 is a rare radioactive isotope that occurs naturally in the atmosphere and in living plants and animals. The half-life of 5,730 years is so low that carbon-14 has not been generally measurable in organic material older than about 40,000 years. Obviously no existing carbon-14 is primordial; the half-life is too short. Instead it is continually being created in the upper atmosphere about 15 kilometers above the Earth's surface as a by-product of cosmic ray bombardment. In the reaction, a nitrogen-14 atom absorbs a neutron, emits a proton, and changes to carbon-14. The newly created carbon is quickly incorporated into carbon dioxide, and thus it is assimilated into the carbon cycle.

Carbon-14 eventually decays back to nitrogen-14. However, the age of carbon-bearing material is not determined from the parent-daughter ratio, but from the ratio of carbon-14 to all other carbon in the sample. Carbon-14, until the advent of thermonuclear explosions, was in equilibrium in the atmosphere; the rate of production was equal to the rate of decay. A plant which removes carbon dioxide from the atmosphere receives a proportional share of carbon-14. When the plant dies, it ceases to absorb carbon dioxide, and with time the proportion of carbon-14 progressively decreases. The carbon-14 method therefore depends on the special assumptions (1) that the rate of carbon-14 production in the upper atmosphere is nearly constant, and (2) that the rate of assimilation of carbon-14 into living organisms is rapid relative to the rate of decay. These assumptions appear to be valid.

Radiocarbon dating is useful for only the last brief portion of geologic time, but a great deal of geologic activity has occurred within this short time span—we did not know exactly *how* much until radiocarbon dates became available. Larger events include the retreat of the last continental ice sheets, accompanying climatic changes, changes in ocean circulation, the postglacial rise in sea level, and the development of human civilization. Radiocarbon has provided an extremely valuable tool to anthropologists and even to historians, as well as to students of recent Earth history. Known historical dates have, in fact, provided excellent checks against the accuracy of the method. Many carbon-bearing substances have been successfully dated, including wood, peat, charcoal, bone, leaves, manuscripts, mummy cloths, rope, and marine shells.

Although the time span concerned is short, it would be difficult to overstate the contribution that radiocarbon dating has made to our understanding of the Earth's recent history. As an example, one of the first things to be dated upon the development of the radiocarbon dating method was the final advance of the Pleistocene continental ice sheet in the Great Lakes region. Prior to radiocarbon dating, the age of this advance could only be guessed by the thickness of the soil developed on the till that was left behind by the ice sheet, and the best guess was about 25,000 years. Carbon-14 dates on a peat bed beneath the till at Two Creeks, Wisconsin, yielded a date of 11,400 years. At once it was possible to infer that a great ice sheet overran a forest in Wisconsin 11,400 years ago, a truly significant bit of information. Adding to the significance was the dating, a short time later, of synchronous equivalents in Europe. Without radiocarbon we could make only a rough guess of the age of the last surge of the Pleistocene continental ice sheet and only an unsupported speculation that it represented a widespread event rather than a local one.

Thorium-230 Method

The dating of marine sediments brought up in long cores from the deep sea poses a big problem. Only the topmost few centimeters are typically within the 40,000-year range of carbon-14. Most of the material of interest is a few hundred thousand years old, and is not usually suitable for dating by any of the methods we have discussed above. The problem has been attacked by using radioactive isotopes produced in the uranium decay series.

Uranium remains largly in solution in the sea, but thorium is quickly precipitated. Thorium-230, produced in uranium 238 series decay in the sea, is incorporated into the ocean floor sediments. Thorium-230 decays with a half-life of 75,000 years. Hence, thorium-230 extracted from sea water, and isolated from uranium, decreases with depth in the sediment. Assuming that both sedimentation rate and thorium-230 precipitation rate at a given place have been constant, in recent geologic history, the concentration of thorium-230 within a long sediment core can be measured relative to the thorium-230 content of the surface layer, and ages of sedimentary layers within the core may be determined. Hence, like the carbon-14 method, the thorium-230 method does not utilize the accumulated radiogenic daughter, but only the amount of parent nuclide that remains. The method is applicable to deep-sea sediments up to several hundred thousand years old.

Thorium-230/Protactinium-231 Method

The ratio of thorium-230/protactinium-231 is another that is useful in dating deep-sea sediments. Protactinium-231 is produced by uranium-235 decay, and like thorium-230, it precipitates quickly in the sea. Decay constants differ by a factor of two, so that the ratio changes regularly with time. Thorium-230 and protactinium-231 come from different uranium isotopes which are intimmately mixed in the oceans, so that it is quite safe to assume that they are everywhere produced in a constant ratio. In principle this method is independent of sedimentation rate, and in this respect it is more versatile than the thorium-230 method. It is, however, useful only for ages up to 150,000 years.

RADIOMETRIC TIME SCALE

If the Phanerozoic portion of the geologic time scale were calibrated in detail, perhaps to stage level, with accurate radiometric dates, then a newly dated rock whether it was igneous, metamorphic, or sedimentary could immediately be placed in the proper stage almost as confidently as if it contained diagnostic fossils. We are still a long way from possessing a quantitative time scale of this accuracy, chiefly because it has proved to be a difficult task to tie radiometric dates to sedimentary rocks whose exact position in the geologic time scale is known. The Phanerozoic time scale is based on numerous sections of sedimentary rocks that have been correlated intercontinentally by means of fossils. Most of our accurate radiometric dates, on the other hand, are from igneous rocks that are difficult to define stratigraphically, and this is the core of the problem.

Today the radiometric time scale is still in its formative stages. Some systems, like the Cretaceous and Tertiary, are quite well dated throughout. Many other parts of the geologic column, however, still contain great gaps that are poorly dated or not dated at all. The date for the Paleozoic-Mesozoic boundary, for example, is extrapolated using sparse Permian and Triassic data from geographically diverse areas. Dates of the system boundaries, as they are presently known, appear adjacent to the geologic time scale on the last page of the book. Some of these are in need of considerable refinement. Future refinements in the radiometric time scale will be brought about chiefly by the same means that have tied radiometric ages into sedimentary strata in the past: by using bracketed igneous intrusives, interbedded volcanic rocks, or authigenic minerals in the strata themselves.

Bracketed Igneous Intrusives

An accurately dated igneous intrusive provides a minimum age for a sedimentary rock that it intrudes and a maximum age for a sedimentary rock that overlies it. Ordinarily the intrusion of igneous material accompanies an orogeny that serves to terminate the preceding sedimentary episode. By the time erosion has run its course and sedimentary conditions are re-established, considerable geologic time has elapsed within which the radiometric date is bracketed. In those rare circumstances where intrusives are bracketed within a small time interval, however, their dates have provided some key points for the radiometric time scale. In the Vosges Massif of eastern France, for example, granites intrude fossiliferous strata of Lower Carboniferous age and are in turn overlain by fossiliferous strata of Middle Carboniferous age. The date of 320 million years for the granite constitutes a key point in the radiometric scale.

Interbedded Volcanic Rocks

Lavas or volcanic ash can be introduced suddenly into sedimentary environments without interruption in sedimentation. Dateable igneous material may thus become interbedded with sedimentary rock dateable by fossils. Lavas can be dated with the whole rock potassium-argon method, and minerals in ash deposits can be dated with both potassium-argon and rubidium-strontium methods. Stratigraphically well-dated sedimentary sequences containing interbedded lavas or ash deposits, although not very common, have already provided some of the most valuable reference points in the radiometric scale.

Dating Sedimentary Rocks Directly

Glauconite, a complex silicate of potassium, aluminium, and iron, is commonly *authigenic* in marine sediment, which means that it actually crystallizes within the environment at the time the sediments accumulate. Authigenic minerals can provide the actual age of the sedimentation and glauconite has provided the majority of direct dates from sedimentary rocks, using the potassium-argon method. Unfortunately, glauconite loses argon if it is buried even to moderate depths. Comparison of glauconite dates with mica and sanidine dates from associated bentonites shows that most Cretaceous glauconite determinations are about 5 percent too young, indicating argon loss. Ages of Paleozoic glauconites tend to be low by 10 to 20 percent. For this reason glauconite dates are usually considered as minimum ages unless substantiated by other methods.

SUBDIVISION OF PRECAMBRIAN TIME

In Precambrian terrains, the igneous and metamorphic rocks provide the vast majority of reliable dates. Igneous and metamorphic episodes commonly affected great areas and these may be used as the basis of major time-stratigraphic units. The Canadian Shield, for example, is divisible into several large provinces which have distinctive rock structure and characteristic radiometric ages (see Fig. 6-9). The ages cluster around 2,480, 1,735, 1,370, and 955 million years, indicating four great orogenic episodes. The oldest orogeny, the Kenoran, was

FIG. 6-9 Major structural provinces of the Canadian Shield and median ages of latest orogenic episode that affected each. (After C. H. Stockwell, 1964.)

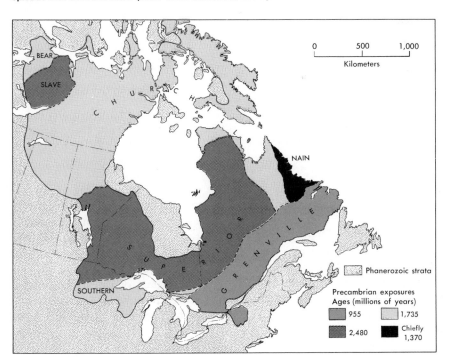

the last period of folding, metamorphism, and plutonic intrusion that affected the Superior and Slave Provinces. The Hudsonian was the last to affect the Churchill, Southern, and Bear Provinces. The Elsonian was the last to affect the Nain Province, and the youngest of all, the Grenville Orogeny, was the last to affect the Grenville Province.

Rocks formed prior to the Kenoran Orogeny, and deformed or recrystallized during the Kenoran, are referred to the Archean Eon. Rocks that are younger are referred to the Proterozoic Eon. The Proterozoic is further subdivided into three eras by the Hudsonian and Grenville Orogenies. The somewhat cumbersome names given to these eras, shown in Fig. 6-10, come from the Greek words meaning past maturity, maturity, and prematurity. Whether or not the names themselves find widespread acceptance, the orogenic episodes on which their classification is based are widely recognizable, on the basis of their distinctive dates, throughout much of North America.

EON	ERA	Major orogenic episodes	Millions of years before present
PROTEROZOIC	Hadrynian		570
		Grenville Orogeny	955
	Helekian — Neohelekian / Paleohelekian	Elsonian Orogeny	1,370
		Hudsonian Orogeny	1,735
	Aphebian		
ARCHEAN		Kenoran Orogeny	2,480

FIG. 6-10 Time-stratigraphic classification of Canadian Shield Precambrian rocks.

Other Precambrian shield regions of the world are made up of similar structural provinces that reflect successive orogenies. In each region the Precambrian is subdivided on local criteria. Precambrian orogenic episodes in some of the other shield regions, although called by different names, appear to coincide in time fairly closely with those on the Canadian Shield. For example, ages of important orogenies on the Australian Shield are 2,300, 1,800 and, 1,400 years—tantalizingly near to those of the Kenoran, Hudsonian, and Elsonian Orogenies. Hence these may have been times of worldwide orogenic episodes. However, much more data will be needed to determine whether or not a worldwide Precambrian classification based on ages of orogenic events will ever be feasible.

AGE OF THE EARTH

We know from extensive sampling of the Earth's crust that today's lead-204, lead-206, and lead-207 exist in the ratios 1.0, 18.5, and 15.6. On the same scale uranium-235 and uranium-238 exist in ratios of 0.0725 and 10.0 The present-day quantities of these nuclides are plotted as the right edge of Fig. 6-11. At any time in the past there must have been more uranium. Knowing the half-lives of uranium-235 and uranium-238 we can calculate how *much* more by simply plotting the decay curves of these isotopes as a function of time.

FIG. 6-11 The radioactive decay of the Earth's uranium has added significant lead-206 and lead-207 and has changed their proportions throughout geologic time. Left margin of lead-207 curve shows that it cannot have been accumulating for more than 5,600 million years, and similarly lead-206 cannot have been accumulating for more than 6,700 million years. Indicated ratios are based on lead-204 = 1.0.

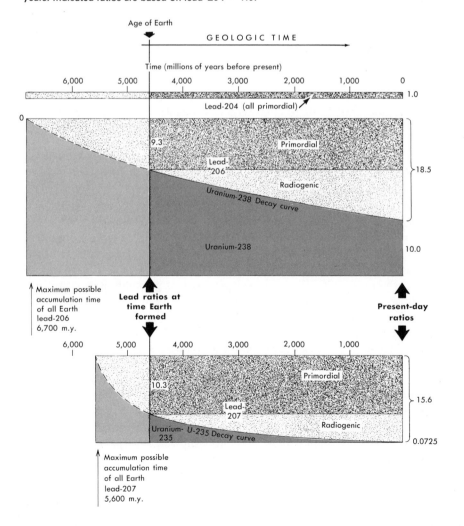

Thus, 4,510 million years ago (the half-life of uranium-238) there was twice as much uranium-238 on Earth as there is today; but at that time there must have been about 80 times as much uranium-235 because 4,510 million years is 6.3 half-lives of uranium-235.

Just as there were greater quantities of uranium-235 and uranium-238 progressively backward in time, there were smaller quantities of lead-207 and lead-206, as the decay curves in Fig. 6-11 show. In fact, if one follows the decreasing quantities of lead far enough to the left in Fig. 6-11 (ignoring for the moment the indicated age of 4,600 million years) one runs out of lead altogether, first out of lead-207 at 5,600 million years and then out of lead-206 at 6,700 million years. This means that the Earth can be no older than 5,600 million years because even if *all* the lead-207 in the world were radiogenic, it would have been produced in the last 5,600 million years. In other words if the Earth were older than this, it would necessarily contain more lead-207 than it does. Similarly all of the lead-206 would have accumulated in 6,800 million years. But there is certainly no reason to suppose that lead-207 and lead-206 began to exist only after the Earth formed. Indeed, the discordance of their maximum possible accumulation times indicated that a portion of the Earth's lead-207 and lead-206 is primordial; that is, it was supplied along with all other nonradiogenic nuclides when the Earth formed.

Thus the Earth must be considerably younger than the 5,600 million years in which all of its lead-207 would have accumulated radiogenically. At the same time it must be older than its oldest rocks. These are gneisses from eastern Greenland, which have rubidium-strontium ages of nearly 4,000 million years. Exactly when within the 5,600 million-year upper limit and the 4,000 million-year lower limit was the Earth born?

As quantities of lead-206 and lead-207 increased through geologic time they progressively changed ratios relative to one another because their parent uranium isotopes decay at different rates. Today there is more lead-206 than lead-207, but 2,000 million years ago they were equally abundant with ratios of about 15.0 (see Fig. 6-11), and prior to 2,000 million years ago there was more lead-207 than lead-206. Thus for each moment of geologic time, quantities of lead-206 and lead-207, measured relative to unchanging lead-204, are unique. If one could establish exactly what these isotopic ratios were at the time the Earth formed, the unique fit on the curves in Fig. 6-11 would determine its age. None of the Earth's rocks preserve primordial lead because since the Earth formed all of its material has been weathered, melted, or otherwise recycled by geologic processes. Thus we have turned to meteorites, rocks which fall from outer space, for samples that have been undisturbed since the time the solar system formed.

Meteorites have a wide variety of compositions: about 7 percent consist mostly of iron, and the remainder are "stones," which consist mostly of silicate minerals. A large proportion of stony meteorites contain peculiar globules of silicate minerals known as "chondrules," and these meteorites are called *chondrites.*

Diamonds occur occasionally in both iron and stony meteorites, indicating that they formed under high pressures and temperatures such as would be attained within a small planet. The crystal structure of many iron meteorites substantiates this conclusion. Yet the composition of some stones suggests rapid cooling. Thus meteorites represent differing modes of origin, but they all formed very early in the history of the solar system and, although they probably fragmented from larger bodies, they have not remelted or recrystallized. They are valuable samples of the primordial stuff of which the solar system was made.

Most stony meteorites contain trace quantities of both lead and uranium. Hence their total lead content, like that of the Earth, is a mixture of radiogenic lead and primordial lead. Iron meteorites and some chondrites, however, contain trace quantities of lead but no uranium. These provide samples of virtually uncontaminated primordial lead, which have isotopic ratios of 1.0 for lead-204, 9.3 for lead-206, and 10.3 for lead-207. These indicate an age of 4,600 million years on the lead evolution curves in Fig. 6-9.

In actual practice the most precise meteorite ages have been obtained using a simple construction called a *lead isochron.* The isochron is a straight line on a graph of lead-206 and lead-207. Its advantage is that a great many points define the line, so that the data are much more reliable than that from a single meteorite alone (see Fig. 6-12). The points represent some meteorites that contain only primordial lead and others that contain varying quantities of radiogenic lead. The *slope* of the isochron changes progressively as the meteorites age.

The dotted *growth curves* on Fig. 6-12 show how isochron dating works. Because the meteorites form from the same cosmic dust at the same time, they initially contain the same lead isotope ratios indicated by the primordial point on Fig. 6-12. A meteorite containing no uranium would, in fact, remain at that point throughout geologic time. However, meteorites that contain uranium increase their content of lead-206 and lead-207 through time by radiogenic additions, and the lead composition of each moves away from the primordial point on a growth curve like those shown. The growth curve for each meteorite at first rises steeply from the primordial lead point while more lead-207 than lead-206 is produced, and then flattens as uranium-235 becomes depleted due to its rapid decay. The greater a meteorite's initial uranium-lead ratio, the greater proportion of radiogenic lead it generates and the higher its curve. At any given time the lead-206 and lead-207 ratios of all the meteorites fall on the same line which passes through the point for primordial lead. This line is the isochron and its slope is a function of the age of the system which the various points represent.

The actual isochron defined by meteorites in Fig. 6-12 has a unique slope for an age of 4,635 million years. The precision in these analyses is considerably

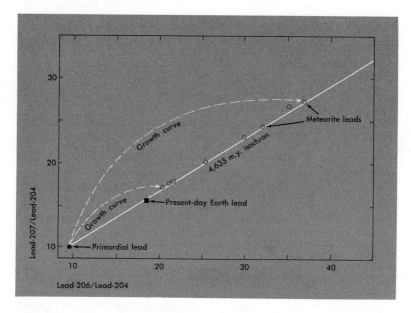

FIG. 6-12 Lead isochron diagram for meteorites gives an age of 4,635 million years. Present-day Earth lead falls on the isochron, indicating that it came from the same primordial source as meteorite lead and at the same time. (After G. R. Tilton, 1973.)

greater than the approximately one percent of uncertainty in the uranium half-lives. The most recent measurement of uranium-235 and uranium-238 half-lives (the most recent is not necessarily the best) would give an age of 4,570 million years for the meteorite isochron. For all practical purposes we may continue to round these determinations to 4,600 million years, a figure that is not likely to be refined significantly in the future.

This may be a convincing age for meteorites, but what right have we to consider this the age of the Earth? Simply this: if the Earth formed earlier or later than meteorites, then its initial lead isotope ratios would have been different (refer again to Fig. 6-9). In this event Earth lead would have evolved from a different primordial point, and its growth curve would not have belonged to the family of growth curves shown in Fig. 6-10. Thus present-day Earth lead would not fall on the isochron for meteorite lead. But Earth lead, analyzed from numerous sources including the oceans and a host of geologically young lead minerals, does, in fact, fall on the meteorite isochron, as Fig. 6-10 indicates. Thus Earth lead also initially had the composition of the primordial point shown. This is excellent evidence that the Earth and meteorites were both isolated from the same cosmic reservoir some 4,600 million years ago.

suggestions for further reading

CHAPTER 1
GROWTH OF THE CONCEPT

ALBRITTON, CLAUDE C., 1967, *Uniformity and simplicity. A symposium on the principle of the Uniformity of Nature*, Geol. Soc. America Special Paper no. 89.

EISELEY, LOREN, 1958, *Darwin's Century*, New York, Doubleday & Co.

————, 1959, Charles Lyell, *Scientific American*, August, 1959, reprint published by W. H. Freeman & Co., San Francisco, 10 pp.

GEIKIE, ARCHIBALD, 1897, *The Founders of Geology*, London, Macmillan and Co.

GOULD, S. J., 1965, Is uniformitarianism necessary? *Am. Jour. Science*, v. 263, p. 223–228.

HARLAND, W. B., A. G. SMITH, and B. WILCOCK (editors), 1964, *The Phanerozoic Time Scale: Part I: Introduction*, Geological Society of London, v. 120 S.

HOLMES, ARTHUR, 1963, Introduction, in *The Precambrian*, v. 1, Kalervo Rankama (editor), New York, Interscience, p. xi–xxiv.

McINTYRE, D. B., 1963, James Hutton and the philosophy of geology, in *The Fabric of Geology*, C. C. Albritton (editor), Reading, Mass., Addison-Wesley, p. 1–11.

TOULMIN, STEPHEN and JUNE GOODFIELD, 1965, *The Discovery of Time*, New York, Harper & Row.

CHAPTER 2
THE ROCK RECORD

DUNBAR, C. O. and J. RODGERS, 1957, *Principles of Stratigraphy*, New York, John Wiley & Sons.

BLATT, H., G. MIDDLETON and R. MURRAY, 1972, *Origin of Sedimentary Rocks*, Englewood Cliffs, New Jersey, Prentice-Hall, Inc.

PETTIJOHN, F. J. and P. E. POTTER, 1964, *Atlas and Glossary of Primary Sedimentary Structures*, Berlin, Springer-Verlag.

SHROCK, R. R., 1948, *Sequence in Layered Rocks*, New York, McGraw-Hill.

REINECK, H., and I. B. SINGH, 1973, *Depositional Sedimentary Environments—With Reference to Terrigenous Clastics*, Berlin, Springer-Verlag.

RIGBY, J. K. and W. K. HAMBLIN (editors), 1972, *Recognition of Ancient Sedimentary Environments*, Society of Economic Paleontologists and Mineralogists, Special Publication no. 16.

SELLEY, RICHARD C., 1970, *Ancient Sedimentary Environments*, Ithaca, New York, Cornell University Press.

CHAPTER 3
TIME STRATIGRAPHY

ADAMS, F. D., 1938, *The Birth and Development of the Geological Sciences*, New York, Dover Publications.

AGER, D. V., 1973, *The Nature of the Stratigraphical Record*, New York–Toronto, John Wiley & Sons.

BERRY, W. B. N., 1968, *Growth of a Prehistoric Time Scale*, San Francisco, W. H. Freeman & Co.

HEDBERG, H. D., 1965, Earth history and the record in the rocks, *Proc. Am. Philosophical Soc.*, v. 109, no. 2, p. 99–104.

MATTHEWS, ROBLEY K., 1974, *Dynamic Stratigraphy*, Englewood Cliffs, New Jersey, Prentice-Hall, Inc.

STORMER, LEIF, 1966, Concepts of stratigraphic classification and terminology, *Earth Science Reviews*, v. 1, p. 5–28.

WELLER, J. M., 1960, *Stratigraphic Principles and Practice*, New York, Harper & Row, particularly Chapter 3, The Geologic Systems.

WELLER, J. M., 1960, Development of paleontology, *Jour. Paleontology*, v. 34, p. 1001–1019.

CHAPTER 4
PHYSICAL CORRELATION AND PALEOGEOGRAPHY

CLOUD, PRESTON (editor), 1970, *Adventures in Earth History*, San Francisco, W. H. Freeman & Co.

COULOMB, JEAN, 1972, *Sea Floor Spreading and Continental Drift*, Dordrecht, Holland, D. Reidel Publishing Co.

COX, ALLAN (editor), 1973, *Plate Tectonics and Geomagnetic Reversals*, San Francisco, W. H. Freeman & Co.

HALLAM, A., 1973, *A Revolution in the Earth Sciences: From Continental Drift to Plate Tectonics*, Oxford, Clarendon Press.

KUMMEL, BERNHARD, 1970, *History of the Earth: An Introduction to Historical Geology*, 2nd ed., San Francisco, W. H. Freeman & Co.

LONGWELL, C. R., Chairman, 1949, Sedimentary facies in geologic history, *Geol. Soc. America* Memoir 39.

MARVIN, R. B., 1973, *Continental Drift: The Evolution of a Concept*, Washington, D.C., Smithsonian Institute Press.

TARLING, D. and M. TARLING, 1971, *Continental Drift: A Study of the Earth's Moving Surface*, Garden City, New York, Doubleday & Co., Inc.

WILSON, J. TUZO, and others, 1972, *Continents Adrift*, San Francisco, W. H. Freeman & Co.

CHAPTER 5
BIOSTRATIGRAPHY

ARKELL, J. W., 1933, *The Jurassic System in Great Britain*, New York, Oxford University Press, p. 14–37.

BERRY, W. B. N., 1966, Zones and zones—with exemplification from Ordovician, *Am. Assoc. Petrol. Geologists Bull.*, v. 50, p. 1487–1500.

CLOUD, P. E., Jr., 1961, *Paleobiogeography of the Marine Realm*, in *Oceanography*, Mary Sears (editor), A.A.A.S. Publication no. 67, p. 151–200.

CRAIG, G. Y., 1966, *Concepts in Paleoecology*, Earth Science Reviews, v. 2, p. 127–155.

HALLAM, A. (editor), 1973, *Atlas of Paleobiogeography*, Amsterdam, London, New York, Elsevier.

ROSS, C. A. (editor), 1974, *Paleogeographic Provinces and Provinciality*, Society of Economic Paleontologists and Mineralogists, Special Publication no. 21.

SHAW, A. B., 1964, *Time in Stratigraphy*, New York, McGraw-Hill.

CHAPTER 6
RADIOMETRIC DATING

FAUL, HENRY, 1966, *Ages of Rocks, Planets and Stars*, New York, McGraw-Hill.

HAMILTON, E. I., 1965, *Applied Geochronology*, London and New York, Academic Press.

HARPER, C. T. (editor), 1973, *Geochronology: Radiometric Dating of Rocks and Minerals*, Stroudsburg, Pennsylvania, Dowden, Hutchinson and Ross, Inc.

HARLAND, W. B., A. G. SMITH, and B. WILCOCK (editors), 1964, *The Phanerozoic Time-Scale; part 2: Radiometric Methods with Respect to the Time Scale*, Geological Society of London, v. 120 S.

HARLAND, W. B., E. H. FRANCIS, and P. EVANS, 1971, *The Phanerozoic Time-Scale: A Supplement*, Geological Society of London, Special Publication no. 5.

PATTERSON, C. C., 1956, *Age of Meteorites and the Earth*, Geochemica et Cosmochemica Acta, v. 10, p. 230–237.

SCHAEFFER, O. A. and J. ZAHRINGER (editors), 1966, *Potassium Argon Dating*, New York, Springer-Verlag.

literature cited

BARRELL, JOSEPH, 1917, Rhythms and the measurement of geologic time, *Geol. Soc. America Bulletin,* v. 28, 745–904.

CREER, K. M., 1965, Paleomagnetic data from the Gondwanic continents, *Phil. Trans. Roy. Soc. London,* series A, no. 1088, v. 258, p. 27–40.

DIETZ, R. S. and J. C. HOLDEN, 1970, The breakup of Pangaea, *Scientific American,* v. 223, October, 1970.

ERICSON, D. B. and GOESTA WOLLIN, 1968, Pleistocene climates and chronology in deep-sea sediments, *Science,* v. 162, p. 1227–1234.

FISHER, D. W., and others, 1962, Geological Map of New York, State Museum and Science Service, Geological Survey, Albany.

HART, S. R., 1964, The petrology and isotopic-mineral age relations of a contact zone in the Front Range, Colorado, *Jour. Geology,* v. 72, p. 493–525.

HEDBERG, HOLLIS D., 1964, Earth history and the record in the rocks, *Proc. Am. Philosophical Soc.,* v. 109, no. 2, p. 99–104.

HURLEY, P. M., and others, 1967, Test of continental drift by comparison of radiometric ages, *Science,* v. 157, p. 495–500.

ISRAELSKY, M. C., 1949, Oscillation chart, *Am. Assoc. Petroleum Geologists Bull.,* v. 33, p. 92–98.

JAMES, H. L., 1960, Problems of stratigraphy and correlation of Precambrian rocks with particular reference to the Lake Superior Region, *Am. Jour. Science,* v. 258-A, p. 104–114.

LARSON, E. E., R. REYNOLDS, and R. HOBLITT, 1973, New virtual and paleomagnetic pole positions from isotopically dated Precambrian rocks in Wyoming, Montana, and Arizona: their significance in establishing a North American apparent wandering path, *Geol. Soc. America Bull.,* v. 84, p. 3231–3248.

LARSON, R. L. and W. C. PITMAN, III, 1972, World-wide correlation of Mesozoic magnetic anomalies, and its implications, *Geol. Soc. America Bull.,* v. 83, p. 3645–3662.

MACKENZIE, D. B., 1963, Dakota Group on west flank of Denver Basin, in *Geology of the Northern Denver Basin and Adjacent Uplifts,* Rocky Mountain Association of Geologists, Denver, p. 135–148.

McCROSSAN, R. G. and R. P. GLAISTER (editors), 1964, *Geological history of eastern Canada,* Calgary, Alberta Soc. Petroleum Geologists.

McDOUGALL, I., 1964, Potassium-Argon ages from lavas of the Hawaiian Islands, *Geol. Soc. America Bull.,* v. 75, p. 107–128.

McKEE, E. D., 1945, Cambrian history of the Grand Canyon region. Part I. *Stratigraphy and ecology of the Grand Canyon Cambrian,* Carnegie Inst. Washington, Pub. 563.

MURTHY, V. RAMA, and others, 1972, Rubidium-strontium and potassium-argon age of lunar sample 15555, *Science,* v. 175, p. 419–421.

OPDYKE, N. D., B. GLASS, J. D. HAYS, and J. FOSTER, 1966, Paleomagnetic study of Antarctic deep-sea cores, *Science,* v. 154, p. 349–357.

O'SULLIVAN, R. B. and E. C. BEAUMONT, 1957, Preliminary geologic map of western San Juan Basin, San Juan and McKinley Counties, New Mexico, U.S. Geol. Survey Oil and Gas Investigations Map OM 190.

PALMER, A. R., 1965, Trilobites of the Late Cambrian Pterocephaliid Biomere in the Great Basin, United States, *U.S. Geol. Survey Prof. Paper 493.*

PEPPER, J. F., and others, 1954, Geology of the Bedford Shale and Berea Sandstone in the Appalachian Basin, *U.S. Geol. Survey Prof. Paper 259.*

PHLEGER, F. B., Jr., 1954, Ecology of foraminifera and associated microorganisms from Mississippi Sound and environs, *Am. Assoc. Petroleum Geologists Bull.,* v. 38, p. 584–647.

RANKIN, D. W., and others, 1969, Zircon ages of felsic volcanic rocks in the upper Precambrian of the Blue Ridge, Appalachian Mountains, *Science,* v. 166, p. 741–742.

REESIDE, J. B., Jr., 1924, Upper Cretaceous and Tertiary Formations of the western part of the San Juan Basin of Colorado and New Mexico, *U.S. Geol. Survey Prof. Paper 134.*

SCHENCK, H. G. and J. J. GRAHAM, 1960, *Science Reports,* Tohoku University, Sendai, Japan, 2nd Ser. (Geol.), Spec. Vol., no. 4, p. 92–99.

143

SCOTT, G. R. and W. A. COBBAN, 1955, Geologic and biostatigraphic map of the Pierre Shale between Jarre Creek and Loveland, Colorado, U.S. Geol. Survey Misc. Geologic Investigations Map I-439.

SLOSS, L. L., 1963, Sequences in the cratonic interior of North America, *Geol. Soc. America Bull.*, v. 74, p. 93–114.

STOCKWELL, C. H., 1964, Fourth report on structural provinces, orogenies, and time-classifications of rocks in the Canadian Precambrian Shield, in *Geol. Survey of Canada Paper 64–17*, pt. II.

TEICHERT, C., 1958, Some biostratigraphical concepts, *Geol. Soc. America Bull.*, v. 69, p. 99–120.

TILTON, G. R., 1973, Isotopic lead ages of chondritic meteorites, *Earth and Planetary Science Letters*, v. 19, p. 321–329.

WELLS, J. W., 1963, Coral growth and geochronometry, *Nature*, v. 197, no. 4871, p. 948–950.

WETHERILL, G. W., 1971, Of time and the moon, *Science*, v. 173, p. 383–392.

WHEELER, H. E., 1964, Baselevel, lithosphere surface, and time-stratigraphy, *Geol. Soc. America Bull.*, v. 75, p. 599–610.

index

Relative Durations of Major Geologic Intervals	Era	Period	Epoch	Duration in Millions of Years (Approx.)	Millions of Years Ago (Approx.)
CENOZOIC		Quaternary	Recent	Approx. last 10,000 years	0
			Pleistocene	2	2
MESOZOIC			Pliocene	3	5
			Miocene	18	23
			Oligocene	15	38
			Eocene	16	54
PALEOZOIC	Cenozoic	Tertiary	Paleocene	11	65
		Cretaceous		71	136
		Jurassic		54	190
	Mesozoic	Triassic		35	225
		Permian		55	280
		Carboniferous — Pennsylvanian		45	325
		Carboniferous — Mississippian		20	345
		Devonian		50	395
		Silurian		35	430
		Ordovician		70	500
	Paleozoic	Cambrian		70	570
PRECAMBRIAN	Precambrian			4,030	4,600

Formation of Earth's crust about 4,600 million years ago

Millions of Years